Distribution Systems Operation and Management

Operational Guide to AWWA Standard G200

G200

Kanwal Oberoi, ME, PE

American Water Works Association

The Authoritative Resource on Safe Water®

Advocacy
Communications
Conferences
Education and Training
Science and Technology
Sections

Operational Guide to AWWA Standard G200
Distribution Systems Operation and Management

Project Manager: Gay Porter DeNileon
Production: Glacier Publishing Services, Inc.

Library of Congress Cataloging-in-Publication Data

Oberoi, Kanwal.
  Operational guide to AWWA standard G200 : distribution systems operation and management / by Kanwal Oberoi.
    p. cm.
    Includes bibliographical references.
  ISBN 978-1-58321-734-4
1. Water--Distribution--Standards--United States. 2. Water utilities--Management--Standards--United States. I. Title.

TD481.O24 2009
628.1'44021873--dc22

                    2009015705

Printed in the United States of America
American Water Works Association
6666 West Quincy Ave.
Denver, CO 80235

Contents

Distribution Systems Operation and Management

SECTION 1: ACKNOWLEDGMENTS

Special thanks are extended to Jaala Draper who provided the technical review for this publication. Also appreciation is extended to Jane Byrne, Patricia Moore, and Kevin Whitsett for their contributions.

Thank you, too, to the committees and volunteers that helped conceive the framework and develop the series of standards that led to the creation of this guidebook, particularly Eva Nieminski, Jim Ginley, and Todd Humphrey for their reviews and passion for excellence in water quality and operations.

SECTION 2: FOREWORD

This operationial guide has been written as a guidance tool for the implementation of water distribution systems operation and management (O&M) best practices, as set forth in ANSI/AWWA G200, *Distribution Systems Operation and Management* (AWWA G200). AWWA G200 is part of the voluntary G-series of standards for water utilities, initially established in 2004.

The original idea for the operational guide series came from utility managers who participated in a two-year pilot project that was conducted and supported by the American Water Works Association (AWWA) Utility Quality Programs Committee, the AWWA Accreditation Committee, and nearly 30 utility professionals from more than 10 North American utilities. The pilot project assessed the applicability and practicality of the series of AWWA Utility Management Standards by working with two utilities—Birmingham (Ala.) Water Works and Sewer Board and Washington County (Va.) Service Authority—and a team of volunteers from utilities, consulting firms, and other organizations.

During the pilot process, the utility managers requested that AWWA develop some type of guidance or "how to" documents to go along with the G-series standards. These guidance documents would serve two purposes: one, to help utility managers understand the purpose and function of these new standards, and two, help them implement and incorporate the standards into everyday operations. From this suggestion the series of operational guides was conceived.

This guidebook is the first of the series that will be developed to correspond with the appropriate G-series standard.

SECTION 3: INTRODUCTION

Just as important as the quality of the water supplying the distribution system is the quality of O&M used to maintain that system. If the infrastructure used to transport water to customers is allowed to deteriorate and operate without any level of standards, the quality of the water delivered cannot be guaranteed safe for consumption, let alone of high quality.

The US Environmental Protection Agency's (USEPA's) Safe Drinking Water Act ensures public safety by setting extensive regulatory requirements for public water suppliers. However, progressive utilities may voluntarily subscribe to additional requirements, such as the AWWA G-series of standards, in order to take a more proactive approach to guarantee that a high-quality product, which exceeds regulatory requirements, reaches the consumers' taps.

ANSI/AWWA G200, *Distribution Systems Operation and Management*, is based on water distribution system O&M best practices, developed by consensus of the AWWA Standards Committee. AWWA G200 also calls for quality O&M of distribution systems through the appropriate expertise of those operating and managing the system. This guidebook has been established to assist any water distribution utility, large or small, in adopting those practices set forth in AWWA G200.

This guidebook breaks AWWA G200 down into the following components:

Standard Language—The Standard Language is wording from a section of AWWA G200, as written in the latest standard document. In this guidebook, the standard language is shaded in gray.

Rationale—The Rationale provides background on the topic in each section of the standard and relevance to why the requirements are important for water distribution utilities.

Example of Methods or Procedures—The Examples given do not cover every aspect of the standard and will not apply to every utility's situation. Additionally, they are not intended to be a complete instruction guide for the implementation of AWWA G200, rather they are intended to point interested parties in the right direction and give insight on how processes and procedures may be properly implemented in accordance with parts of the standard. In order to maintain this publication at a reasonable size, only a handful of examples per topic are included, and most point to a secondary publication with current standard operating procedures and practices used by a handful of select utilities in a standardized format. There are also many sources of additional information listed in the References and Resources section (Sec. 6).

Questions to Check Progress—The Questions are listed as a tool to self-assess the status of the utility's compliance with AWWA G200. If the utility can confirm that they have all items in the questions in place, they may wish to consider applying for recognition.

The Audit Checklist in Sec. 7 is included to further assist utilities in performing an initial gap analysis or assessing their progress in implementing AWWA G200. This tool will help identify any gaps that may exist in their current procedures and those required to meet the standard. It asks specific questions and looks for proof and documentation that procedures are indeed in place where called for by the standard. Just as the examples are not an instruction manual for implementing all of AWWA G200, neither is the checklist a blueprint. However, a utility that has all the components represented in the checklist is likely to have a majority of the G200 standard well covered in their operating and management procedures and practices.

SECTION 4: REQUIREMENTS

Sec. 4.1   Water Quality

4.1.1   **Compliance With Regulatory Requirements**

The utility shall determine and document all local, state, provincial, federal, or other regulations that apply to their utility. The utility shall meet or exceed all applicable regulations.

Rationale

Customers associate health and safety issues related to drinking water directly to the utility's ability to meet all local, state, provincial, federal, or other regulatory requirements to which the utility subscribes. It is absolutely necessary for the utility to meet or exceed all applicable regulations to gain the confidence of the consuming customers. To comply with these requirements, the utility shall establish and maintain procedures to identify and have access to the applicable legal and regulatory requirements to which the utility subscribes related to its scope and operation. The utility shall ensure that these applicable legal and regulatory requirements are taken into account in establishing, implementing, and maintaining its water quality management system.

This can be accomplished by the utility developing a legal and other regulatory requirements list and ensuring this list is periodically reviewed and updated. The utility shall also document how these requirements are adhered to in its operation.

Example of Methods or Procedures

Following is a sample of a regulatory and other requirements form that can be used by a utility to identify and maintain a listing of the most up-to-date versions of all regulatory and other requirements that apply to their system.

Title: Regulatory and Other Requirements Form

Purpose/Scope: This document is used to identify and record legal and other requirements to which the utility subscribes that have been identified as being directly applicable to the utility's activities, products, and services. Completion of this document will aid overall management of the water quality.

Instructions: This document (Figure 4.1.1) is to be updated as regulatory and other requirements are modified, added, or removed. List the title of the

Prepared By/Date:			
Title of Rule, Regulation, or Law	General Subject Covered	Responsibility for Current Version	Document Format/ Location (If Electronic, Provide Web Address)

Figure 4.1.1   Regulatory and Other Requirements Form

requirement and provide a brief description in the column titled: General Subject Covered. Rules, regulations, and laws must be periodically reviewed for current version and updates; therefore, identify which part of the utility will perform this review in the column titled: Responsibility for Current Version. Establish electronic links in the column titled: Document Format/Location. This ensures that the most current requirements are available for reference. With hard-copy requirements, list the location(s) of the hard-copy.

All utility departments are required to ensure that regulatory and other requirements are interpreted and applied to their operations. Those listed as "Responsible" are required to periodically ensure that the most recent legal and other requirements are identified and listed. NOTE: This form may be replicated on a computer or duplicated on a photocopier. The computer copy must look similar to this document and contain the same information.

Questions to Check Progress

1. Does the utility have an established list of updated legal and regulatory requirements?

2. Has the utility communicated this information to all the stakeholders and established systems to periodically review and update this form?

3. Has the utility established programs to ensure regulatory requirements are adhered to?

4. Has the utility established a system of monitoring to ensure compliance to the legal and regulatory requirements and to ensure water quality in the distribution system?

4.1.2   Monitoring and Control

4.1.2.1   Sampling plan. The utility shall establish a routine distribution system sampling plan that is representative of the entire distribution system. The sampling plan shall be reviewed annually and adjustments made based on historical data trends, changes in water use patterns, or other changes that may affect water quality. The utility shall analyze data trends and have an action plan to respond to changes.

4.1.2.2   Sample sites. Sample sites shall include, at a minimum, sites required for regulatory compliance monitoring. Additional sites shall be sampled as necessary to provide a complete picture of water quality in the system. The utility shall use sampling sites that are representative of all known distribution system conditions, including the following: variations in hydraulic detention time, pipeline materials, where booster disinfection is applied, where water is stored, and where water quality deterioration (e.g., loss of disinfectant residual and increased microbial growth) is known, or suspected, to occur. Samples shall include locations that represent the longest detention time in the system, dead-end locations, areas of low circulation, and finished water storage facilities. Locations where problems have occurred in the past require more frequent sampling.

4.1.2.3   Sample collection. Samples shall be collected in accordance with *Standard Methods for the Examination of Water and Wastewater*. Chain of custody forms shall be used throughout the sampling process in accordance with *Standard Methods for the Examination of Water and Wastewater* or regulatory requirements. Both sample collectors and the laboratory shall use standardized labels and forms.

4.1.2.4   Sample taps. Sample taps shall be protected from outside sources of contamination. The integrity of the sample taps shall be evaluated annually to correct leaks or other potential sources of contamination.

Rationale

Water quality monitoring in the distribution system provides information on water quality conditions between the entry point to the system and the customer's tap. It guides the utility in making changes in operations, maintenance, or treatment to improve overall water quality before water reaches the customer. The most common monitoring parameters used for distribution system water quality include coliform bacteria; chlorine residual; heterotrophic plate count (HPC) bacteria; pH; turbidity; disinfection by-products (DBPs); color, taste and odor; pressure; and water temperature.

Sampling for water quality within the distribution system is required by law and a sampling plan will ensure that the utility will meet or exceed all sampling requirements when it is properly established. Selecting appropriate sample sites throughout the distribution system will ensure that the samples will be representative of the entire system and that water quality conditions are acceptable throughout. It is important to determine locations in the system where water quality is more likely to be problematic and sample regularly at those sites. These locations include, but are not limited to, dead ends, areas of low circulation, and extreme outlying points in the distribution system, where disinfectant residual is more likely to be low and poor water quality conditions could easily develop. Sampling at these predetermined locations at regular intervals will ensure that the water quality is being maintained there and throughout the rest of the system.

Written procedures for proper field sampling, sample collection, and sample handling will ensure that all sampling results will be accurate and truly representative of the distribution system. All associates who may be expected to perform sample collection and/or analysis need to be fully trained on these procedures to ensure sample integrity and results.

Routine inspection of all sample taps is critical to obtaining accurate water quality samples. Sample taps can be inspected continuously as they are accessed for sampling, but annual inspections should be conducted at a minimum to ensure that they are still accessible, safe from contamination, and in good condition.

Example of Methods or Procedures

Following is a method for water quality sampling procedures, instructing associates on the proper sampling, field testing, handling, and processing procedures to follow. Associates who are responsible for performing any sampling and/or field testing must be trained on these procedures.

Title: Water Quality Sampling

Purpose and Scope: To provide guidance, steps, and instructions on field testing and collection of water quality samples in compliance with the Safe Drinking Water Act for all types of water quality parameters within the distribution system.

Responsibility and Authority: All associates performing water quality testing and/or sampling shall fully and completely ensure that all samples are collected in compliance with the Safe Drinking Water Act. The water quality technician will ensure proper maintenance and annual review of the sample site plan, proper calibration of all field testing units, and availability of current forms.

Work Steps: Water Quality Field Testing—Turbidity Level Testing

1. Collect a sample from a hydrant or faucet in a clean container. Fill sample bottle to the 15-mL line.

2. Wipe the sample bottle with a soft, lint-free cloth to remove water spots and fingerprints. Always use clean sample bottles.

3. Apply one drop of silicone oil to the outside of the sample bottle and wipe with a soft cloth to obtain an even film over bottle.

4. Turn the turbidity meter instrument on at the "I/O" switch.

5. Make sure that the instrument is on a flat, sturdy surface.

6. Allow air bubbles to settle out of the sample bottle.

7. Place sample bottle in the instrument so that the diamond mark aligns with the raised mark in front of the compartment.

8. Press "READ" and the instrument will display "NTU." Record the turbidity after the lamp turns off.

   - If a sample causes an unstable reading, hit the "SIGNAL AVERAGE" button and then the "READ" button again, which will measure the sample 10 times within ±20 seconds.

   - Signal averaging uses more power and should only be used under unstable reading conditions.

9. When turbidity testing is complete, clean all equipment, flush out sample bottle with distilled water, and secure equipment within the case.

10. Log data on appropriate form and/or database.

One way to ensure sample tap integrity is to institute sample tap inspections into the daily duties of the associate(s) responsible for performing routine sampling of the distribution system. Each sample tap can be inspected every time it is visited for sample collection.

Questions to Check Progress

1. Does the utility have a representative distribution system water quality sampling plan that meets or exceeds the regulatory requirements?

2. Does the utility review and update the water quality sampling plan annually, or earlier when major changes occur in the distribution system, to ensure it remains representative of the distribution system?

3. Does the utility have written sampling standard operating instructions to meet the requirements of the standard's methods and local regulatory agencies in time of collection, transportation, and chain of custody requirements?

4. Does the utility identify, install, and maintain the sampling locations at least annually to ensure that those sites are prevented from possible contamination?

4.1.3	**Disinfectant Residual Maintenance**

(Sec. 4.1.3 does not apply to distribution systems not utilizing a residual disinfectant.)

4.1.3.1   Disinfectant residual. The utility shall maintain a detectable disinfectant residual or a heterotrophic bacteria (or plate) count (HPC) of 500 or fewer colony forming units per mL at all points in the distribution system at all times. The utility shall monitor and record disinfectant residual or HPC as described in the sampling plan (Sec. 4.1.2.1).

4.1.3.2   Nitrification control. If the utility adds ammonia to the water as part of the disinfection process or if there is a significant concentration of ammonia in the source water, the following procedures shall be used:

a. Monitoring of the free ammonia concentration prior to and after chloramination. The weight ratio of chlorine to ammonia must be monitored and controlled to minimize the presence of free ammonia in the system.

b. Adjustment of the ammonia feed rate to compensate for the source water ammonia concentration, or the ammonia is oxidized prior to chloramination. Establish and periodically review a disinfectant goal based on the historical database to avoid nitrification.

4.1.3.2.2   The utility shall routinely monitor for other key nitrification indicator parameters (i.e., nitrite, nitrate, and free ammonia).

4.1.3.3   Booster disinfection. (Sec. 4.1.3.3 does not apply to distribution systems not utilizing a residual disinfectant or that do not employ booster disinfection.)

4.1.3.3.1   The utility shall set and document residual goals and a program to monitor compliance with goals as detailed in Sec. 4.1.3.1.

4.1.3.3.2   The utility shall have clear operating procedures for each booster disinfection facility based on maintaining residual goals at critical points. Plans should take into account seasonal variations, water quality, flow, and system operation variations.

4.1.3.3.3   The utility shall have a written plan to respond to any variance between operational goals and actual measured results.

4.1.3.4   Disinfection by-product monitoring and control.

4.1.3.4.1   The utility shall have a program to monitor and control disinfection by-products. The program shall establish goals for DBPs at critical points in the distribution system.

4.1.3.4.2   The utility shall have an action plan to respond to DBP problems in the distribution system. This plan shall include specific actions to take should the DBP levels exceed the established goals.

Rationale

The disinfection of drinking water has been hailed as one of the most important advances ever in the protection of public health. In the United States and other developed countries, drinking water disinfection has all but eliminated serious waterborne diseases that still kill tens of millions of people each year in other parts of the world.

In North America, chlorine residual is probably the second most important parameter in the distribution system. In addition to its regulatory importance, its function as a surrogate for the condition and health of the distribution system is widely held. For example, chlorine residual information is helpful in identifying areas of the distribution system with a high disinfectant demand due to long hydraulic detention times or microbial regrowth problems. Chlorine, in some form, is the only disinfectant known that can provide a continuing level of disinfection (known as *secondary* or *residual disinfection*) after water has left the treatment plant and enters the distribution system. HPC bacteria monitoring provides a general measure of bacterial water quality in the distribution system. It may be used in place of disinfectant residual to meet certain Surface Water Treatment Rule (SWTR) requirements.

All disinfectants create a complex family of by-product compounds, some of which are of human health concern, and are now regulated by USEPA. For example, both gaseous chlorine and sodium hypochlorite react with organic material in water to form chlorinated organic compounds. In response to this, many drinking water utilities in the United States are implementing chloramination for distribution system disinfection to minimize formation of the two major types of DBPs, trihalomethanes (THMs) and haloacetic acids (HAAs). However, there is a significant concern associated with chloramination—nitrification. Nitrification is caused by the growth of ammonia-oxidizing bacteria (AOB) in the distribution system. AOB use ammonia as an energy source, converting ammonia to nitrite. Nitrite speeds the decomposition of the chloramine residual, the disappearance of which can increase HPC bacteria and coliforms. Reduced residual levels also may cause a utility to violate disinfectant residual requirements.

Most of the studies of DBPs in drinking water over the past 30 years have been conducted in laboratories under controlled, uniform conditions. Little is known about changes in DBP concentrations and speciation that can take place in actual distribution systems, where DBP formation can be affected by pipe walls, biofilms, flow rates, residence time, changing storage tank levels, booster chlorination, and free chlorine residuals. That is why careful monitoring around areas of known variances, such as booster chlorination stations, is so critical.

Every utility must balance their use of disinfection chemicals with the formation of harmful DBPs. Monitoring and control plans must be established and implemented in order to do so.

Example of Methods or Procedures

Example 1

Standard procedures can include guidelines for chlorine residual monitoring. The following is one example of a standard procedure that includes chlorine residual monitoring.

Title: Chlorine Residual Monitoring

Chlorine Level Testing

- Zero out the chlorine meter (NOTE: Zeroing must be done prior to every test):
    1. Fill a clean sample bottle to the 10-mL mark with the same solution as being tested.
    2. Wipe excess liquid off sample cells or damage to the instrument may occur.
    3. Place the water-filled sample in the cell compartment with the diamond mark facing the front of the instrument. Cover the sample bottle with the instrument cap.
    4. Ensure the cap is placed over the sample with the curved surface facing the keyboard.
    5. Press the "ZERO" key.
    6. After approximately two (2) seconds, the display will read "0.00."
- Using the Hach Pocket Colorimeter (NOTE: Samples must be analyzed immediately on-site):

1. Fill a 10-mL sample bottle with the sample to be tested.
2. Add the contents of two DPD Total Chlorine Powder Pillows to the sample bottle. Cap and gently shake for 20 seconds.
3. Wipe excess liquid off sample cells or damage to the instrument may occur.
4. Wait three minutes, then place the sample bottle into the cell holder. NOTE: A pink color will form if chlorine is present.
5. Tightly cover the cell with the instrument cap.
6. Press "READ."
7. The instrument will show "___ mg/L" total chlorine.

Work Completion:

1. Chlorine readings within the water distribution system should be between 1.0 and 3.0. If levels obtained are not within this range, then flushing may be required. Contact the senior manager for direction.
2. When chlorine testing is complete, clean all equipment, flush out sample cell with distilled water, and secure equipment within the case.
3. Log data on appropriate form and/or database (Figure 4.1.3).

In addition to regulatory sampling of the system, chlorine residuals should be routinely monitored at dead ends and other trouble areas that have been identified throughout the system.

Example 2

The following procedure is an example of how one system attempts to control nitrification. This does require steps to be taken by other parts of the system besides distribution, as it is a whole-system approach.

			Chlorine Residual		Turbidity		
Location (address)	Size (in.) [cm]	Date (mm/dd/yy)	Pre-flush	Post-flush	Pre-flush	Post-flush	Consumption (gal) [m³]
						Total Consumption:	0

Annual Blow-Off Monitoring Schedule
_____ Quarter

Figure 4.1.3   Annual Blow-Off Monitoring Schedule Form

Title: Nitrification Prevention

Purpose: Outline protocols and chemical treatments to minimize the potential for nitrification in the distribution system

Responsibility and Authority: Process Control/Shift Supervisors

Work Steps:

1. Control chlorine-to-ammonia ratio in 3:1 to 4:1 range. If ratio exceeds 5:1, notify Director/Assistant Director. Total chlorine and ammonia fed to Booker and Stoney is displayed on SCADA—also check monthly chemical inventories for usage as well as log sheets for daily feed.

2. If SCADA Hi or Hi Hi alarm goes off based on Chemscan free ammonia sensor, check that chlorine and ammonia feed systems are operating correctly—notify Director/Assistant Director.

3. Chlorine dioxide, as well as being an excellent disinfectant, produces chlorite as a by-product. Chlorite has been shown to inhibit growth of bacteria that cause nitrification. Ensure chlorine does not exceed the maximum contaminant level (MCL) as it is regulated.

4. Monitor turnover of storage tanks. Typically 3–4 days to completely turn a tank is adequate, but during hot, wet weather, more aggressive tank turning may be necessary (>60 percent per day).

5. Have lab take samples and measure nitrite/nitrate on weekly basis, especially during hot, wet weather when tank turnover may be an issue.

6. Monitor combined chlorine levels in tanks—if it starts to trend down from a normal range of 2–2.5, call Plant Director.

7. Monitor combined chlorine throughout system—if it falls below 0.5 (verify by grab), call Plant Director and distribution personnel to initiate flushing.

8. Coordinate with Planning Deptartment when tank cleaning task instructions are issued (typically 3–5 year cycle).

Example 3

AWWA M20, *Water Chlorination/Chloramination Practices and Principles*, recommends three different strategies for booster, or secondary, chlorination for utilities with chloraminated water.

- Add free chlorine to combine with residual ammonia. The procedure uses residual free ammonia remaining in the water as a consequence of chloramine decay. Chlorine is added in the ratio of 4.5:1 to 5:1 (chlorine:

ammonia-nitrogen ratio) and a chloramine residual is reformed. This process has an added benefit because free ammonia is combined and, therefore, is not available to cause nitrification problems.

- Add chlorine and ammonia. Chloramine booster stations are rare, but they have been used in situations where there is not enough free ammonia remaining to reform sufficient chloramine concentrations for residual disinfection. Some stations use sodium hypochlorite and aqua ammonia to form the chloramine.

- Add chlorine to breakpoint. This strategy is employed when a free chlorine residual is desired. Chlorine is added to water that contains residual ammonia and/or a small amount of chloramine. The amount added is sufficient to surpass the breakpoint and achieve a free chlorine residual. This process eliminates any nitrification potential due to the presence of free ammonia. This procedure is used mostly where there is a separation between two systems or where an area of one system can be isolated as long as THMs are not a problem in the area to be kept on free chlorine. One system or area can maintain a free chlorine residual and the other a chloramine residual.

Questions to Check Progress

1. Does the utility have documented and implemented systems and procedures to maintain the chlorine residual or a heterotrophic plate event of 500 or fewer colony-forming units per milliliter at all points of the distribution system as per the local regulatory requirements?

2. Does the utility have documented and implemented systems and procedures to monitor and control the formation of nitrification? Does this include corrective actions to be taken when the nitrification takes place in the distribution system?

3. Does the utility have a chlorine residual maintenance goal and a written procedure or booster system to maintain chlorine residual in the distribution system if needed?

4. Does the utility have a documented program to monitor, control, and take corrective action that includes the action for the DBP at the critical points in the distribution system?

4.1.4   **Additional Requirements for Utilities Not Utilizing a Disinfectant Residual**

The utility shall monitor and record heterotrophic plate count (HPC).

4.1.4.1   Response program. The utility shall have an action plan to respond in the event the HPC goal is not met in the distribution system. The action plan shall include

1. Goals for HPC at critical points in the distribution system.

2. Criteria for initiation of actions defined in the plan.

3. Criteria for initiation of actions that correct a problem before it becomes a health or regulatory concern.

4. Description of specific responsibilities of staff and assignment of responsibilities.

Rationale

There are three main types of microorganisms that can be found in drinking water—bacteria, viruses, and protozoa. These can exist naturally or can occur as a result of contamination from human or animal waste. Some of these are capable of causing illness in humans. The main goal of drinking water treatment is to remove or kill these organisms to reduce the risk of illness.

The HPC test is a method for monitoring the overall bacteriological quality of drinking water. In distribution systems, HPC provides some indication of stagnation, tuberculation, residual disinfectant concentration, and availability of nutrients for bacterial growth. In addition, in chloraminated systems, HPC can reflect improper chlorine/ammonia ratios or a problem with nitrification. Where the HPC is 500/mL or less, the water is deemed to have a detectable residual for compliance with USEPA's Surface Water Treatment Rule. Therefore, any system not monitoring disinfectant residual must monitor HPC throughout the distribution system, in order to meet regulatory requirements and provide safe drinking water to the service population.

Example of Methods or Procedures

In the United States, HPC monitoring can substitute for disinfection residual monitoring for systems that do not use a disinfectant residual. Monitoring and response plans for HPCs should mirror those for disinfection residual monitoring.

Questions to Check Progress

1. Does the utility have an implemented system and documented procedures to monitor, control, or take corrective action to maintain the HPC at the critical points in the distribution system?

2. Does the utility have a goal and action plan that contains the goal for HPCs at the critical criteria for initiation of an action plan that corrects the problems?

3. Does the utility drill annually to ensure proper implementation of the action plan?

4. Are all the staff with maintenance, control, and corrective action responsibilities trained on an annual basis?

4.1.5	**Internal Corrosion Monitoring and Control**
4.1.5.1	Prevention and response program. The utility shall have an action plan to respond to internal corrosion and deposition problems in the distribution system. The action plan, at a minimum, shall include

1. Monitoring and sampling plan for corrosion-related parameters (i.e., pH, alkalinity, conductivity, phosphates, silicates, calcium, metals, asbestos, etc.). The scale-forming potential shall be measured by the appropriate indices (calcium carbonate precipitation potential [CCPP] or the Langelier saturation index [LSI] or some other set of parameters proven to protect the system).

2. Inspection, when exposed, of the condition of piping for perforations, tuberculation, and other conditions related to structural integrity and hydraulic capacity.

3. Procedures to control lead and copper levels.

4. Guidelines for controlling other corrosion related by-products, such as iron, color, zinc, and taste and odor, in the distribution system.

Rationale

The greatest concern for internal corrosion of water distribution system piping is the introduction of harmful substances to the water supply, particularly lead and copper. Another major concern is the build up of scale from corrosion products on the interior pipe walls. These tubercles and scale on the interior of the pipes can result in clogs, restricting flows and furthering the loss of structural integrity.

The Lead and Copper Rule (LCR) was added to the Safe Drinking Water Act in 1991. This rule requires systems to monitor drinking water at customer taps. If lead and/or copper concentrations exceed their action levels, the system must undertake a number of additional actions to control corrosion. These requirements can include additional water quality parameter monitoring, corrosion control treatment, source water monitoring/treatment, public education, and lead service line replacement. High levels of lead and/or copper discovered in water obtained through customers' taps is a good indication of corrosive water. However, by monitoring pH, alkalinity, conductivity, calcium, and other corrosion-related parameters, a system should be able to identify any issues with the corrosiveness of the water before there is a health threat present. All large systems are required by the LCR to perform regular water quality parameter monitoring and to provide corrosion control treatment.

Even with a monitoring plan and developed action plan for containing internal corrosion, this chemical process will still occur over time, requiring the rehabilitation and replacement of infrastructure. The structural integrity of the mains should be monitored whenever an opportunity presents itself.

Example of Methods or Procedures

Figure 4.1.5 is an example of a sampling plan for monitoring corrosion-related parameters—in accordance with the lead and copper rule. In addition to the sampling schedule, there are set limits on point-of-entry and water quality (POE and WQ) parameters to ensure corrosion control is maintained.

Lead and Copper Rule Schedule

*WQP = field pH, field temp, alkalinity, conductivity, calcium, orthophosphate
**POE = pH and orthophosphate from finished water

Period	Year	WQP* Period	WQP* # Sites	Send to State	POE** Period	POE** # Sites	Send to State	Tier 1 Pb 7 Cu Period	Tier 1 Pb 7 Cu # Sites	Send to State
Reduced	2003	1st Qtr	10	Quarterly	Twice/mo (minimum)	1	Quarterly	Jun–Sept	50	by Oct 10th
		2nd Qtr	10	Quarterly						
		3rd Qtr	10	Quarterly						
		4th Qtr	10	Quarterly						
Triennial	2004	1st Qtr	10	Quarterly	Twice/mo (minimum)	1	Quarterly			
		2nd Qtr	3	Annual						
		3rd Qtr	3	Annual						
		4th Qtr	2	Annual						
Triennial	2005	1st Qtr	2	Annual	Twice/mo (minimum)	1	Quarterly	Jun–Sept (any summer within the 3-year period)	50	by Oct 10th
		2nd Qtr	3	Annual						
		3rd Qtr	3	Annual						
		4th Qtr	2	Annual						
Triennial	2006	1st Qtr	2	Annual	Twice/mo (minimum)	1	Quarterly			
		2nd Qtr	3	Annual						
		3rd Qtr	3	Annual						
		4th Qtr	2	Annual						
Triennial	2007	1st Qtr	2	Annual	Twice/mo (minimum)	1	Quarterly			
		2nd Qtr	3	Annual						
		3rd Qtr	3	Annual						
		4th Qtr	2	Annual						
Triennial	2008	1st Qtr	2	Annual	Twice/mo (minimum)	1	Quarterly	Jun–Sept (any summer within the 3-year period)	50	by Oct 10th
		2nd Qtr	3	Annual						
		3rd Qtr	3	Annual						
		4th Qtr	2	Annual						
Triennial	2009	1st Qtr	2	Annual	Twice/mo (minimum)	1	Quarterly			
		2nd Qtr	3	Annual						
		3rd Qtr	3	Annual						
		4th Qtr	2	Annual						

NOTE: This schedule assumes the lead and copper levels remain below the action levels. If the lead and copper levels are above the action level, increased frequency and sites are required. See *Federal Register* for details.

Figure 4.1.5   Sampling Plan: Lead and Copper

Questions to Check Progress

1. Does the utility have an implemented and documented program for the internal corrosion and deposition monitoring and control in the distribution system?

2. Does the utility use treatment technology to ensure that internal corrosion and deposition are effectively controlled and maintained in the distribution system?

3. Does the utility effectively monitor and measure the water quality parameters such as pH, alkalinity, conductivity, phosphates, silicates, calcium, metals, etc., and measure the calcium carbonate precipitation potential (CCPP), the Langelier saturation index (LSI), or some other

parameters on a regular basis to ensure internal corrosion and deposition in the distribution system is effectively controlled?

4. Does the utility have documented procedures that are implemented to control lead and copper along with other corrosion by-products such as iron, color, zinc, and taste and odor in the distribution system?

4.1.6   Aesthetic Water Quality Parameters

4.1.6.1   Color and staining. The utility shall have an action plan to address color and staining problems. The action plan, at a minimum, shall include

1. An inquiry call system in place that can differentiate between color and staining problems and other inquiries, and track them.

2. Trained personnel who can handle customer inquiry calls over the phone, can explain system problems that are known, and can collect pertinent information for response personnel.

3. Communication of inquiry information to a response team for a timely resolution. Review of inquiry records for data trends to identify problem areas of the distribution system.

4.1.6.2   Taste and odor. The utility shall have an action plan to address taste and odor problems. The action plan, at a minimum, shall include

1. An inquiry call system in place that can differentiate between taste and odor problems and other inquiries and track them.

2. Trained personnel who can handle customer inquiry calls, can explain system problems that are known, and can collect pertinent information for response personnel.

3. Communication of inquiry information to a response team for a timely resolution. Review of inquiry records for data trends to identify problem areas of the distribution system.

Rationale

National Primary Drinking Water Regulations are legally enforceable standards that protect public health by limiting the levels of contaminants in drinking water. National Secondary Drinking Water Regulations are nonenforceable guidelines regulating contaminants that may cause negative aesthetic effects (such as taste, odor, or color) in drinking water. USEPA recommends secondary standards to water systems but does not require systems to comply. However, it is the aesthetic water quality conditions that are noticeable to the public.

Aesthetic complaints include taste, odor, color, and clarity. These aesthetic qualities will influence consumer confidence and perception of the water's safety. Sensory properties of drinking water are the first observation of a problem by consumers. Though many of the unusual aesthetic properties detected do not have health effects, they could be indicative of serious safety and health concerns that may otherwise go unnoticed. For example, a pungent hydrocarbon odor may indicate a petroleum contamination, or an excessively salty taste in the water may prove to be an overdose of fluoride or sewage contamination.

The absence of a universal detector for all possible contaminates and the high cost involved to install real-time physiochemical parameter monitoring devices at all points throughout the distribution system make having a call and tracking system for these types of inquiries essential for utilities to protect public health, as well as perception. It is essential to obtain pertinent information from the affected party and respond immediately in order to discover and rectify any serious safety and health issues. All concern calls of this nature should be tracked and analyzed so that any patterns indicating a serious or chronic water quality issue can be identified and remedied.

For the above reasons, utilities should take customer aesthetic water quality concerns seriously and have procedures and trained personnel in place to respond to, resolve, and track such concerns.

Example of Methods or Procedures

The following example is a procedure to follow when a customer calls in any type of water quality concern. It guarantees that the concern is investigated, resolved, and tracked, as well as ensures that the water is safe for the public.

Title: Water Quality Reporting/Investigation

Purpose and Scope: To provide guidance, steps, and instructions in the investigation and recording of customer concerns regarding water quality, color, particles, taste, and odor. All water quality customer concerns will be channeled through the dispatcher's office. The preventive maintenance section will contact the customer and determine the cause of the problem, perform field tests and sampling, and determine the corrective actions needed.

Responsibility and Authority: The investigator will ensure that the concern is resolved fully, or followup steps are being taken to resolve the concern.

Work Preparation: All customer water quality concern calls must be forwarded to the dispatcher's office. Any water quality concern call received by an associate must be forwarded to the dispatcher on duty via telephone contact.

The dispatcher will create a service request and submit it to the distribution department for investigation through the customer management system. The dispatcher will include the following information in the service request:

1. Caller's name, phone number, and address.
2. Account number, meter number.
3. Detailed description of the problem (color, particles, odor, taste, cloudy, milky, etc.).
4. Length of time experiencing the problem.
5. Does the problem occur with hot water, cold water, or both?
6. Has problem been noticed throughout the house/building?
7. Does the customer use a "whole house" or "point-of-use" filter system?
8. Has the customer tried flushing their system to attempt to clear up the problem? If not, ask the customer to try flushing. (Flushing at the bathtub, for a higher flow rate, is recommended if the problem is throughout the entire house/building.)

Pre-Trip: If possible, call the customer to set up an appointment and interview the customer to assess the nature of the complaint. Determine if anyone has been sick to assess whether or not any potential health problems exist.

Work Steps: Meet with the customer and explain what procedures will be used to investigate their concern. Information to obtain from the customer:

1. What is the nature of the problem?
2. How long has the problem persisted?
3. When was the problem first noticed?
4. What if anything has changed in the household?
5. Does the problem only appear at certain times of the day or during certain activities?
6. Is the problem unique to the customer and residence, or has the problem been noticed by neighbors, friends or relatives, or at their workplace?
7. What sort of dwelling is it? Is there anything special about the location or nearby buildings?
8. Does the building use hot water space heating?
9. Are there any possibilities of a cross connection?
10. Has there been any construction in the area?

11. Have hydrants been flushed in the area?
12. Does a lawn sprinkler system or fire sprinkler system exist?
13. Is there a water treatment or filtration device present?
14. Are drink machines, ice makers, air conditioners, water softeners, or any unusual devices connected to the water supply?

Tasks to Perform:

- Conduct field chlorine and turbidity tests.
- If either test does not meet set standards, begin system flushing.
- If sickness or contamination is suspected, collect a bacteriological sample or contact the technician for water quality to assist sampling (if in doubt, collect a bacteriological sample).
- If no one is at home or additional samples are required, such as ammonia, metals, lead, etc., contact the water quality technician to coordinate an appointment with the customer.
- Perform additional tests as necessary. If results still do not meet standards, test a different water service or hydrant.
- If the standards are still not met, a possible distribution system contamination exists. Immediately inform the distribution systems engineering manager.

Work Completion: Clean up the work site, and clean and put up tools used.

Advise the customer of what was done, the test results obtained, and the status of their water quality. If no problem can be determined, then advise the customer that the problem may be in their internal plumbing system and they can contact a plumber to investigate further.

Deliver any samples to a certified lab.

Questions to Check Progress

1. Does the utility have a system with documented procedures to address the aesthetic quality of the water in the distribution system? Are these systems and procedures implemented as a preventive maintenance program?

2. Do the procedures include a system that tracks all the customer concerns from the initiation of the concern to proper resolution to the satisfaction of the customers? Is documentation kept to meet or exceed the local regulatory requirements?

3. Does the utility have trained personnel with expertise to resolve these concerns through proper identification, scientific testing, and interpretation? Do personnel receive training at least on an annual basis?

4. Does the utility have a goal to reduce these concerns as a continuous improvement?

4.1.7 Customer Relations

4.1.7.1 Customer inquiries. The utility shall have a system to document customer inquiries.

1. The system shall record the customer identification, specific inquiry type, the result of the investigation, and the resolution of the inquiry.

2. The system shall document the number of customer inquiries.

3. The utility shall establish an annual goal to continually reduce the number of water quality complaints and track trends.

4. The utility goal shall be to maintain or reduce customer complaints. Customer inquiries shall receive an initial contact by the investigator within 8 hours of the receipt of the inquiry.

5. The utility shall evaluate customer inquiries related to water quality (i.e., taste and odor, color or staining, corrosion, etc.) and shall implement a plan to investigate and respond to inquiries.

6. The system shall have trained personnel who can handle customer inquiries, can explain system problems that are known, and can collect pertinent information for response personnel.

7. The system shall communicate the inquiry information to a response team for a timely resolution.

8. The system shall review inquiry records for data trends to identify problem areas of the distribution system.

4.1.7.2 Service interruptions. The utility shall have a system to document all planned and unplanned service interruptions. The utility shall have an annual goal to continually reduce unplanned service interruptions. As an example, the goal may be based on miles of pipe or number of taps.

Rationale

The public relations of a utility should be directed at maintaining public confidence and customer satisfaction. Utilities may not have to deal with market competition; however, satisfied customers are more tolerant of temporary service disruptions or water quality issues when they occur. Also, customers who are happy with their utility's management and operations are more understanding of necessary rate increases.

Call-center personnel and water distribution field staff have the greatest exposure to the public and therefore the greatest chance of positively or negatively influencing the public's perception of the utility. There are several techniques that can be implemented to make sure that this influence is positive, and not negative. One such technique is by documenting customer concerns and inquires so that they can be recorded for later review and trending and responded to in a timely manner without being overlooked.

Service interruptions are a necessary part of maintaining a water distribution system. However, this inconvenience to customers has a major impact on public perception and confidence in the utility. It should be every system's goal to minimize service interruptions, planned and unplanned. Not only does tracking the number and types of service interruptions give an indication of customer satisfaction, it can also give the utility a better idea of system integrity. High numbers of unplanned service interruptions may expose an unknown operational problem.

Example of Methods or Procedures

Example 1

Figure 4.1.7 is an example of a water quality complaint form, detailing the type of information that should be collected from a concerned caller. This form could be adapted for a variety of concern calls that may typically be received.

Name/Address/ Business/Subdivision	Date:	Water Quality Complaint				Grid#:	Assigned to
	Time:	Color			Milky	Grid#:	
	Tel #:	Taste			Other		
	Acct #:	Odor			Particles in Water	Main Size:	
		Pressure			Air in Lines		
	Meter #:	Illness			Knocking in Pipes	DE/LP	Work Request #:
		Leak			Rust Remover		
	Description:						

Figure 4.1.7   Water Quality Complaint Form

Example 2

One way that a system can monitor service interruptions is to track the pertinent information through their customer management system. One utility asks the following questions on all O&M type work orders:

1. Was water service interrupted?
2. Was the interruption planned or unplanned?
3. How many customers were without service?
4. How long did the interruption to service last?

Questions to Check Progress

1. Does the utility have a system in place to document customer information, inquiry type, and results and resolution when an inquiry is made?

2. Has the utility established an annual goal to maintain or reduce the number of water quality complaints?

3. Are inquiry records reviewed periodically for trends to identify problem areas in the system?

4. Does the utility have a system to document all planned and unplanned service disruptions and a goal to continually reduce unplanned service interruptions?

4.1.8   **System Flushing**

The utility shall develop and implement a systematic flushing program that meets the needs of the utility, taking into consideration the condition of the system, hydraulic capacity, treatment, water quality, and other site-specific criteria. At a minimum, the flushing program shall incorporate the following items:

1. The program addresses a preventive approach to distribution system flushing, including occasional spot flushing to address localized problems or customer concerns and routine flushing to avoid water quality problems.

2. The utility shall perform system flushing at the velocity appropriate to address water quality concerns.

3. The utility has written procedures addressing all activities associated with system flushing, water quality, monitoring, frequency, locations, and duration, as well as adherence to all regulatory requirements.

Rationale

Distribution systems are primarily designed for fire flow. Because of this, normal operating velocities are low, resulting in the accumulation of sediment in water mains. For that reason, a systematic flushing program is recommended to maintain water quality to the customer. Additionally, purging stagnant water and maintaining adequate chlorine residual in the distribution system will keep the drinking water bacteriologically safe.

An effective flushing program is one that anticipates and prevents water quality problems and customer complaints. The characteristics of an effective flushing program are as unique as the individual distribution system and several methods can be employed; however, a minimum velocity of 2.5 ft/s (0.8 m/s) is required. The two general methods are conventional and unidirectional flushing.

- Conventional Flushing: This method can be used when responding to customer complaint calls or in areas of the distribution system that experience chronic water quality problems. It can either be in the form of a systematic program where all hydrants are flushed or only those that are considered critical in maintaining adequate water quality.

- Unidirectional Flushing: This method begins at a known "clean" source and expands throughout the distribution system. By isolating valves to create a "one-way" flow, the velocity in the water main is increased, removing sediment and biofilm. This method can also incorporate valve exercising, hydrant maintenance, and hydraulic testing.

Example of Methods or Procedures

The following procedure is for the implementation of a systematic flushing program. This procedure encompasses a comprehensive approach to preventive maintenance (PM), in addition to flushing requirements.

Title: Unidirectional Flushing Program

Purpose and Scope: To provide guidance, steps, and instructions when performing unidirectional flushing (UDF) operations. The UDF program consists of the combination of hydrant PM, valve PM, system flushing, and hydraulic testing.

Responsibility and Authority: The UDF foreman shall ensure that all hydrants and valves are fully and completely operated and inspected, mains are flushed properly, and all discovered deficiencies are properly noted and logged. Also, the foreman will ensure that flooding resulting in property damage does not occur. In addition, the foreman will ensure that the residual pressure does not fall below 20 psi (138 kPa), and turbidity levels are 5 ntu or less, per State Primary Drinking Water Regulations.

Work Preparation: Acquire a copy of the scheduled Unidirectional Flushing Run Worksheet, which includes the location of the "run," directions, and a map showing the location of valves, hydrants, mains, etc.

Notify the water treatment plant process control before any flushing is performed in the area of the booster pump stations.

Set up "lane closed" signs, cones/barricades, construction signs, etc., and make the jobsite safe for workers, pedestrians, and motorists, as per the state DOT Traffic Control Handbook.

Set up "flooded area" signs 200 ft (61 m) in each direction of the hydrant to be flushed to warn motorists as they approach the area.

Notify all high-risk customers who may be affected by a possible pressure loss, i.e., schools, hospitals, dialysis customers, and any industry that depends on constant water pressure for normal production.

Do a preliminary check of the run by driving through the area to

1. Locate and check all isolation and additional valves within the run for accessibility, and verify accuracy of valve card information. If there is no valve card on a valve in the run, field sketch the valve locations.

2. Check and clear all plugged or stopped-up catch basins, drainage ditches, etc.

3. Inspect the flow area for any low-lying or flood-prone conflicts that could cause damage to property.

4. Check for any traffic obstructions.

5. Look for construction in the area that could affect the run.

Work Steps: Using the run diagram on the worksheet, operate all additional valves on the worksheet.

1. Attach a static pressure gauge setup to the gauged hydrant, leaving the other nozzle caps on. Open the hydrant, bleed off the air from the valve on the gauge setup, take a static pressure reading, and record the reading in the appropriate column on the worksheet.

2. Install a flushing diffuser equipped with a pitot gauge on the discharge hydrant.

3. Open the discharge hydrant all the way and take the pressure reading. Log the psi reading in the appropriate flow column on the UDF Run Worksheet.

4. At the gauged hydrant, record the pressure reading while the discharge hydrant is flowing. Look for the most constant reading to log. This reading is called the *residual pressure* and should be logged on the worksheet in the appropriate column. At this time, the discharge hydrant should flow for only as long as it takes to read and log the residual and flow pressure (approximately one minute or less).

5. After the static, residual, and flow pressures have been logged on the worksheet, shut down both the gauged and discharge hydrants. Remove the static gauge apparatus from the gauged hydrant. Leave the pitot diffuser attached to the discharge hydrant as it will be used during the UDF run.

6. Using the provided grids and valve cards, list the hydrant number, location, corner/side, barrel type, manufacturing code, valve opening, year, and model in the provided columns on the UDF Run Worksheet for the gauged hydrant, discharge hydrant, and all additional hydrants.

7. Perform preventive maintenance on all the hydrants within the run. On the additional hydrants not flushed, flush for one minute through one 2½-in. nozzle diffuser to clear the water from the hydrant leads and to

get flow readings. Log collected information in the appropriate columns on the worksheet.

8. Operate and close all the isolation valves and operate all the additional valves marked for the run. The isolation valves will be circled on the run diagram.

9. List all required valve maintenance information for both the isolation valves and additional valves in the appropriate columns on the worksheet and fill out a work request form to be turned in to the senior distribution technician.

10. Open the discharge hydrant all the way and start timing the flow with a stopwatch. Take the pressure reading and log it in the appropriate flow column on the worksheet. This is the UDF flow and should be logged in the bottom half of the appropriate flow box.

11. Recheck the area to ensure no flooding, safety hazard, or damage can occur from flowing the hydrant.

12. Begin checking the turbidity level at the discharge hydrant using the turbidity meter.

13. The desired turbidity reading to complete a successful UDF run is 1.5 ntu or less. Once the desired turbidity level is achieved, shut down the discharge hydrant. Turn off the stopwatch and log the flush time and turbidity reading in the appropriate columns on the worksheet. Do not allow flow time to exceed one hour. If the desired ntu level is not achieved within this time, turn off the stopwatch and log the flush time and turbidity reading in the appropriate columns on the worksheet.

14. Reopen the isolation valves and log collected information on the worksheet.

15. When flushing is completed in the area of the booster pump stations for the day, call water treatment plant process control and let them know.

Work Completion:

1. Pick up cones, signs, barricades, etc.

2. Log the hydrant and valve deficiencies found during the run in the appropriate columns provided on the UDF Run Worksheet.

3. Note any drawing, valve, or hydrant discrepancies on the worksheet and on a copy of the valve card, if available.

4. Turn in all valve card corrections, UDF Run Worksheets, and field sketches to your supervisor at the end of the day.

Questions to Check Progress

1. Does the utility have an implemented and documented system and procedures to document and address the customer concerns in a timely manner? Does the system receive calls from customers on a 24-hour, 7-days-a-week basis?

2. Is the system comprehensive enough to document at least customer identification, inquiry type, results of investigation, and resolution of inquiry?

3. Does the utility document all the inquiries and establish a goal to reduce these concerns as per continuous improvement goals for the utility? Does the utility have contact with the customer during the time the concern is reported or respond to the customer within eight hours to confirm the receipt of the inquiry and either its progress or resolution?

4. Does the utility have an implemented and documented system to respond to all planned and unplanned service interruptions, along with a goal to reduce these interruptions based on some specific criteria?

Sec. 4.2   Distribution System Management Programs

4.2.1	**System Pressure**
4.2.1.1	Minimum residual pressure. The minimum residual pressure at the service connection under all operating conditions shall not be less than 20 psi (138 kPa).
4.2.1.2	Pressure monitoring. Pressure shall be monitored at key locations. Pressure alarms may be used to alert operators of pressure conditions outside the utility requirements.

Rationale

Low pressures in the distribution system can result not only in insufficient fire-fighting capacity but can also constitute a major health concern resulting from potential intrusion of contaminants from the surrounding external environment. A minimum residual pressure of 20 psi (138 kPa) under all operating conditions and at all locations (including at the system extremities) should be maintained to help ensure public safety and health.

Example of Methods or Procedures

Example 1

When low-pressure concerns are initiated by customers, follow the initiation steps outlined in the Water Quality Reporting/Investigation procedure discussed in Sec. 4.1.6. Then, once at the customer's home:

1. Unscrew the screens in the taps to determine if they are plugged.
2. Find the tap closest to the meter and screw a pressure gauge to the tap.
3. Open the tap to see what the static pressure is and compare it to hydrant reading.
4. If pressure is within 5 psi (34 kPa) of what it should be for the area, then the problem may be volume and not pressure.
5. Check the type of pipe; galvanized or lead pipe may collect deposits and become narrow.
6. Have the customer open a tap and flush a toilet, then check the pressure gauge while the toilet is running to see what the pressure drop is.
   - If the drop is less than 5–7 psi (34–48 kPa), no problem exists. Explain findings to the customer.
   - If the drop is in excess of 7 psi (48 kPa), check the stop and waste (S&W) valve and meter valves for fully open position.
7. Return to the premises and turn on a tap to see if the flow has improved at all.
8. If nothing has changed, have the customer make sure the fixtures are closed and, using a sonophone, listen to the pipes, fixtures, and meter for hissing noises indicating a leak. Return to the meter and check the leak indicator dial (small red dial) to see if it is moving. If it is, then the leak is private.
9. If hissing is detected, then close the meter valve to see if the hissing stops.
   - If the hissing stops, then there is a leak on private property.
   - If the hissing continues, then there is a leak on the utility side of the meter.
10. If the leak is on private property, then advise the customer of the leak and inform them that they should contact a plumber and are responsible for the repair to be completed.

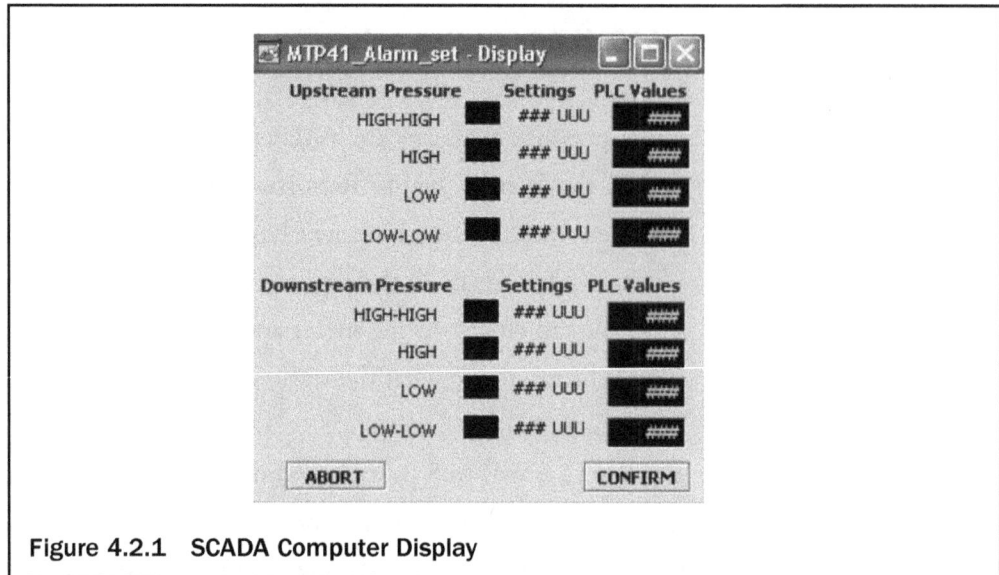

Figure 4.2.1   SCADA Computer Display

Example 2

Most utilities use a supervisory control and data acquisition (SCADA) system to continuously monitor pressure at key points throughout the distribution system. Alarm set points can be set to sound audible and visual alarms when excessively high or low pressures are reached. Figure 4.2.1 is a screen shot of how set points can be entered into a particular SCADA system.

Questions to Check Progress

1. Does the utility have written policies and procedures to monitor and maintain the pressure in the distribution system to meet the local plumbing and regulatory requirements?

2. Does the utility have policy and written procedures to ensure system pressure does not fall below 20 psi (138 kPa) at the service connection?

3. Does the utility have written documented procedures to take corrective action as per the local regulatory requirements if the residual pressure in the distribution system falls below 20 psi (138 kPa)?

4. Does the utility continuously monitor distribution pressure at key locations in the distribution system with alarms to ensure that adequate pressure is maintained throughout the distribution system?

4.2.2   **Backflow Prevention**

The utility shall have a comprehensive cross-connection control and backflow-prevention program as required by federal, state, provincial, and local regulations and at least as stringent as those provided in AWWA Manual M14, *Recommended Practice for Backflow Prevention and Cross-Connection Control.*

Rationale

Cross-connection control is one of the most important barriers in the multiple-barrier approach drinking water suppliers use to protect public health. Contamination of a drinking water distribution system through a cross-connection often results in immediate adverse health effects—illness or even death. Cross-connections are frequently created by individuals who are not familiar with the hazards, even though they may otherwise be well trained and experienced. It is the utility's responsibility to protect public health, even if the hazard originates on the customers' premises.

The water purveyor shall ensure that effective backflow-prevention measures commensurate with the degree of hazard are implemented to ensure continual protection of the water in the public water distribution system. To reduce the risk private plumbing systems pose to the public water system, the water purveyor's backflow-prevention program should include public education regarding the hazards backflow presents to the safety of drinking water.

Example of Methods or Procedures

Title: **Approved Backflow-Prevention Assembly Tester Practices & Code of Conduct**

As per the Water Rules and Regulations, Section X, the Utility has established a cross-connection control program for the benefit of its customers. The program is maintained to ensure that the Utility exercises its responsibility to deliver safe drinking water to these customers. This program requires certain customers to install backflow-prevention assemblies that need testing on a regular basis when notified by the Utility. The Utility requires anyone wishing to test these backflow-prevention assemblies for our water users within our jurisdiction to maintain, as a minimum, a State tester certification. In addition, the Utility will require adherence to all cross-connection control program rules and policies, which are maintained and available for downloading at www.Utility.com and included here by reference. In addition to the cross-connection control program rules and policies, the Utility will now require all utility-approved testers to adhere to the following Tester Practices and Code of Conduct. If a tester is found out of compliance with any cross-connection control program rules, policies, or these tester practices, then they may be denied inclusion on the Utility's List of Certified Testers and, as a result, any submitted test reports would not be recognized or accepted. The Utility's Tester Practices and Code of Conduct are as given below:

1. The Approved Tester shall only record data and sign test forms of assemblies they have tested. A tester shall not falsify any data or results obtained from the field test and reported on the Field Test and Maintenance Form. All data shall be legibly submitted.

2. The Approved Tester shall not make any *unnecessary* repairs and/or replacement of parts or assemblies.

3. When backflow-prevention assembly repairs are necessary, the Approved Tester shall only use original factory repair parts as manufactured for the assembly.

4. The Approved Tester shall not circumvent the installed backflow-prevention equipment by removal or alteration of the assembly or any of its components. In extreme cases, such as severe freeze damage and when no immediate replacement is available, this may be allowable if approval by the Utility is granted and will be determined on a case-by-case basis.

5. The Approved Tester shall perform only utility-specified field-test procedures on a backflow-prevention assembly. A copy of the test procedures and needed equipment are available on the Utility's Web site. When testing an assembly, the tester shall ensure that all components and accessories of the assembly are present for field-testing. Any discrepancies shall be listed in the comments section of the Field Test and Maintenance Report.

6. The Approved Tester shall submit the appropriate passing or failing Field Test and Maintenance Report to the Utility within seven calendar days of performing the field test. If a test report is received after this time period, the test will not be accepted and the tester will be required to retest the assembly in question at their own expense and submit the test form within seven days of the subsequent field test. The tester shall provide a copy of the completed Field Test and Maintenance Report to the water user on completion of the field-test procedures.

7. The Approved Tester shall only test backflow-prevention assemblies with a properly working test kit. The tester shall only use equipment that is in proper working order and within factory accuracy specifications. This test kit shall be checked for accuracy at least once a year or when in need of calibration. A copy of the annual calibration report and any subsequent reports of each test kit shall be submitted to the Utility for our records. Field Test and Maintenance Reports submitted using differential pressure gauges that have not been checked for accuracy within one year or are out of factory accuracy specifications will not be accepted by the Utility and will be returned to the tester.

8. The Approved Tester shall confirm that *all* data (make, model, size, serial number, meter number, etc.) on the Field Test and Maintenance Form is correct for the assembly being tested.

9. The Approved Tester shall observe existing installations of backflow-prevention assemblies to ensure that the assembly is the correct type of protection for the degree of hazard present and properly installed as per the Utility's installation requirements. If the assembly is not the correct type of protection, or installation requirements are not proper, the tester shall proceed with the field test and note the discrepancy on the Field Test and Maintenance Report.

10. The Approved Tester shall comply with the Utility's cross-connection control officials in the exercise of their duties as covered by the rules and practices. The tester shall supply any requested information and/or appear at the site of the assembly when requested.

11. The Approved Tester has neither the responsibility nor the authority to represent or enforce the Utility's cross-connection control program. Enforcement lies solely with the Utility and its appointed officials.

12. All new installations and/or change outs of existing backflow-prevention assemblies shall be installed in conformance with the Utility's Cross-Connection Control Program Manual.

13. The Approved Tester *may not* withhold submission of the Field Test and Maintenance Report to the Utility until payment is made by the customer to the tester. While it is understood that some customers are late in making payment for tester services, or in the past have refused to make payment, the small claims court is the appropriate avenue to pursue compensation. Testers failing to submit test results based on these

circumstances will be subject to removal from the List of Approved Testers. Once you have performed the test, you are *required* to submit your passing or failing Field Test and Maintenance Report to the administrative authority within the prescribed time period and without exception.

14. All Approved Testers desiring to remain on our List of Approved Testers shall participate in the recertification process *at the Utility's recertification site only* at least once every three years—*no exceptions.* Additionally, Approved Testers will have their test equipment examined at least once every three years (differential pressure gauge with calibration certificate, short tube, vertical sight tube, and compensating tee/bleed valve) to ensure that the equipment meets industry, state, and utility standards, rules, and policies.

Any Approved Tester failing to comply with the Utility's Tester Practices & Code of Conduct and all cross-connection control program rules and policies shall be subjected to loss of recognition of test reports and removed from the list of testers within the Utility's area of authority. The Utility will establish a notice system of noncompliance to Tester Practices & Code of Conduct and cross-connection control rules and policies that shall operate as follows:

- First Offense—Verbal warning and consultation with a compliance inspector or cross-connection manager.
- Second Offense—Written warning and/or suspension of testing privileges for up to 30 days within the Utility's distribution system, depending on the frequency and severity of the offense.
- Third Offense—Revocation of testing privileges within the Utility's distribution system.

NOTE: If a tester is found to have falsified a Field Test and Maintenance Report, a legal and defensible document, they will be revoked from testing privileges within our service area. Requests for reinstatement will be reviewed on a case-by-case basis.

The Utility's Required Field Test Procedures, Program Ordinance, Program Manual, Field Test and Maintenance Report, Clarification Statements, and other related documents are available for downloading in Adobe Acrobat Reader format at www.utility.com/. Go to the link titled "Developers/Contractors" for the link to the cross-connection page. *Periodic updates are made to our Web site and it shall be the responsibility of all Approved Testers to visit our Web site to ensure they have the most current material or documents.*

A copy of the Clarification Statement is included with this Policy.

CC-023 Tester Ethics and Code of Conduct—Rev 6-5-2007.doc		
Tester Affidavit: Please complete all information and return this page to the Commission of Public Works within 10 days.		
I hereby certify that I have thoroughly read and understand the above verbiage and agree to fully conform to the provisions of this Backflow-Prevention Assembly Tester Practices & Code of Conduct Policy.		
PLEASE PRINT LEGIBLY:		
Tester Name	State Certification Number	
Company Name		
Company Address		
City	State	Zip Code
Home Address		
City	State	Zip Code
Home Telephone	Work Telephone	
Fax Number	Pager	Cell Phone
Signature:		Date
Email Address:		
NOTE: This document will be maintained and archived by the Utility. Failure to return this affidavit portion will be construed as an Approved Tester's unwillingness to conform to these provisions. Accordingly, that Tester's name will be removed from our List of Approved Testers.		
Received by Utility Inspector:		Date

Questions to Check Progress

1. Does the utility have a comprehensive cross-connection control program as identified by AWWA Manual M14?

2. Does the program also comply with the local plumbing and regulatory requirements?

3. Does the utility have cross-connection devices installed, maintained, and tested as per AWWA Manual M14 on all service connections?

4. Does the utility conduct water quality sampling to identify and document permeation for the susceptible parameters if applicable in the distribution system?

4.2.3   Permeation Prevention

If the utility has permeable (plastic) components (seals, pipes, valves, etc.) that may be susceptible to external contamination (solvents, gasoline, and other organic contaminants), the issue shall be addressed in the utility operation plan.

Rationale

There are many documented instances of permeation of petroleum hydrocarbons into water lines that have impacted public drinking water. A variety of pipe materials are especially vulnerable to permeation. Both polybutylene (PB) and polyethylene (PE) are easily susceptible to permeation. Polyvinyl chloride (PVC) pipe is also vulnerable. Volatile organic compounds (VOCs) have even permeated newly installed asbestos–cement pipe materials. Joint gaskets also have a high permeability that should be considered even though their mass transfer area is relatively minor. Gaskets are also easily chemically degraded by petroleum hydrocarbons. There are instances where gaskets have become so degraded by contaminants that the waterlines have failed. It is important to note that pipe wall thickness will not prevent permeation of VOCs through PB, PE, and PVC pipe material. It is also important to realize that once breakthrough has occurred, the problem will not be solved through flushing. Once compromised, permeated plastic piping must be replaced because the piping will retain its swollen porous state after permeation and chemical degradation.

Example of Methods or Procedures

There are many ways that a utility can provide a measure of permeation prevention for the distribution system. One way to try to avoid this issue is to not allow any PVC piping to be used in construction. Also a utility may specify extra precautions for areas of the distribution system that are within a certain radius of a potential contamination source, such as gas stations.

Questions to Check Progress

1. Does the utility have policies and implemented and documented procedures to prevent permeation in the distribution system?

2. Does the utility have implemented and documented procedures to monitor and take corrective action for the control of the permeation in the distribution system?

3. Does the utility have procedures to document the location where the distribution system is susceptible to the permeation and conduct monitoring for the control of permeation at these locations?

4. Does the utility conduct water quality sampling to identify and document permeation for the susceptible parameters applicable in the distribution system?

4.2.4   Water Losses

4.2.4.1   Water loss. The utility shall have an annual goal for the amount of water loss. The utility shall have documentation defining what is included in this calculation.

4.2.4.2   Response program. The utility shall have an action plan to respond if the annual goal is not met.

4.2.4.3   Leakage. The utility shall have a system for estimating (quantifying) leakage on an annual basis. The system shall express leakage in terms of gal/day/mi ($m^3/d/km$) of distribution pipe.

Rationale

Inefficiencies in water distribution systems associated with underground water system leakage result in a major loss of revenue. Increases in pumping, treatment, and operational costs make these losses prohibitive.

Awareness that water loss is occurring in a water system is the first step in identifying leaks and making repairs. Once water loss has been documented and identified, a water system operator can then determine whether the water loss is a real loss or an unavoidable loss. The first step in accounting for water used and lost in a water distribution system is appropriate data collection.

In general, a 10 to 20 percent allowance for unaccounted-for water is normal. But a loss of more than 20 percent requires priority attention and corrective actions. However, advances in technologies and expertise should make it possible to reduce losses and unaccounted-for water to less than 10 percent. While percentages are great for guidelines, a more meaningful measure is volume of lost water. Once the volume is known, revenue losses can be determined and cost-effectiveness of implementing corrective action can then be determined.

Example of Methods or Procedures

Figure 4.2.4 is an example of a form used to determine the amount of water that is being lost throughout the system without being accounted for. This form provides a good estimation of where all the treated water is going. Information for converting to gpm (gallons per minute) in 2½ in. and 4½ in. fire flow nozzles is shown in Tables 4.2.4-1 and 4.2.4-2.

Water Distribution				
Preventive Maintenance	Senior Dist. Tech. - PM			
Unidirectional Flushing:				1,649,747
Customer Concerns:				979,135
Blow-off Sampling:				4,169,635
Hollywood Fountain (metered–not billed)				90
Fire Department				
Hydrant Testing:				6,717,164
Training:				966,000
Fire Fighting:				42,500
Maintenance		No. of	Main	Consumption
Main Breaks:	Dist. Supervisors	Breaks	Size	Per Month
		1	1"	2,400
		0	1-½"	0
		0	2"	0
		1	2-¼"	7,200
		0	3"	0
		6	4"	141,120
		2	6"	108,000
		2	8"	188,160
		0	10"	0
		2	12"	422,880
		0	16"	0
		0	20"	0
		0	24"	0
		0	30"	0
		0	36"	0
		0	48"	0
			Total:	869,760
Meter Technology	Dist. Supervisor—Meter Technology			
Annual % Error—Underregistering:				2.3%
Technical Support	Tech. Services Supervisor			
Leak Detection: (Except main breaks)				0
Miscellaneous Losses				
City Street Sweeping:				26,000
N. City Street Sweeping:				8,833
WW Collection Usage:				25,000
Cross-Connection	Cross-Connection Manager			
Permitted Hydrant Users:				8,688,000
Unmetered Fire Service Testing:				43,750
Design and Construction	Construction Manager			
Technical Services, Hydro Testing:				111,740
Construction Sites:				250,000
Miscellaneous and Special Projects:				3,643,200
Grand Total Losses/Month:				**28.2 MG**

Figure 4.2.4   Unaccounted-for Water Losses Form

Table 4.2.4-1   Fire Flow Conversion Chart Pitot Gauge Conversion From psi to gpm

2½ in. Nozzle

psi	gpm	psi	gpm	psi	gpm	psi	gpm
1	168	26	856	51	1,198	76	1,463
2	237	27	872	52	1,210	77	1,472
3	291	28	888	53	1,222	78	1,482
4	336	29	904	54	1,233	79	1,491
5	375	30	919	55	1,244	80	1,501
6	411	31	934	56	1,256	81	1,510
7	444	32	949	57	1,267	82	1,519
8	475	33	964	58	1,278	83	1,529
9	503	34	978	59	1,289	84	1,538
10	531	35	993	60	1,300	85	1,547
11	557	36	1,007	61	1,311	86	1,556
12	581	37	1,021	62	1,321	87	1,565
13	605	38	1,034	63	1,332	88	1,574
14	628	39	1,048	64	1,342	89	1,583
15	659	40	1,061	65	1,353	90	1,592
16	671	41	1,074	66	1,363	91	1,601
17	692	42	1,087	67	1,373	92	1,609
18	712	43	1,100	68	1,384	93	1,618
19	731	44	1,113	69	1,394	94	1,627
20	750	45	1,126	70	1,404	95	1,635
21	769	46	1,138	71	1,414	96	1,644
22	787	47	1,150	72	1,424	97	1,653
23	805	48	1,163	73	1,434	98	1,661
24	822	49	1,175	74	1,443	99	1,670
25	839	50	1,187	75	1,453	100	1,678

Table 4.2.4-2   Fire Flow Conversion Chart Pitot Gauge Conversion From psi to gpm

4½ in. Nozzle

psi	gpm	psi	gpm	psi	gpm	psi	gpm
0.25	270	5.25	1,250	10.25	1,740	18.5	2,350
0.5	390	5.5	1,280	10.5	1,760	19	2,380
0.75	470	5.75	1,310	10.75	1,790	19.5	2,410
1	550	6	1,330	11	1,810	20	2,440
1.25	610	6.25	1,360	11.25	1,830	21	2,500
1.5	670	6.5	1,390	11.5	1,850	22	2,560
1.75	720	6.75	1,420	11.75	1,870	23	2,610
2	770	7	1,440	12	1,890	24	2,670
2.25	820	7.25	1,470	12.5	1,930	25	2,720
2.5	860	7.5	1,490	13	1,970	26	2,780
2.75	900	7.75	1,520	13.5	2,000	27	2,830
3	940	8	1,540	14	2,040	28	2,880
3.25	980	8.25	1,570	14.5	2,080	29	2,940
3.5	1,020	8.5	1,590	15	2,110	30	2,990
3.75	1,050	8.75	1,610	15.5	2,150	31	3,030
4	1,090	9	1,630	16	2,180	32	3,080
4.25	1,120	9.25	1,660	16.5	2,210	33	3,130
4.5	1,160	9.5	1,680	17	2,250	34	3,170
4.75	1,190	9.75	1,700	17.5	2,280	35	3,220
5	1,220	10	1,720	18	2,310	36	3,270

Questions to Check Progress

1. Does the utility have implemented and documented policies and procedures for water loss control?

2. Does the utility monitor the unaccounted-for water as per AWWA Manual M36 and have a goal to reduce it to an acceptable level with time?

3. Does the utility have a comprehensive systemwide leak survey and water audit program to identify and repair hidden leaks in the distribution system?

4. Does the utility have a proactive basis to address the concerns of water loss control in the distribution system, along with customers' internal plumbing system, by education and awareness?

4.2.5   **Valve Exercising and Replacement**

4.2.5.1   Valve exercising program. The utility shall have a valve exercising program. This program shall include at least the following elements:

    a. A goal for the number of transmission valves to be exercised annually based on the percentage of the total valves in the system.

    b. A goal for the number of distribution valves to be exercised annually.

    c. Measures to verify that the goals are met and written procedures for action if the goals are not attained.

    d. Critical valves in the distribution system shall be identified for exercising on a regular basis. Potential quality and isolation concerns shall be recognized. The program shall track the annual results and set goals to reduce the percent of inoperable valves.

Rationale

Aging water distribution systems affect water supply reliability. An effective distribution system management program should include a valve exercising program that includes both static and dynamic data, such as

Static Data:
- Identifier Number
- Specific Location
- Make
- Size
- Depth

Dynamic/Maintenance Data:
- Number of Turns
- Operating Condition
- Valve Box Condition
- Open/Closed

A major step in successfully implementing a valve exercising program is setting reasonable goals based on the distribution system size and available resources.

Example of Methods or Procedures

Title: Air Release Valve Inspection Detailed

Purpose and Scope: To provide guidance, steps, and instructions in the detailed inspection of all air release valves (ARVs). These are guidelines for the proper operation, maintenance, and inspection of ARVs.

Responsibility and Authority: The distribution systems supervisor or valve operations foreman shall fully and completely ensure that the ARV is in the open position and that advanced corrosion is not apparent before leaving the work site.

Work Preparation: This activity may be initiated by deficiencies found during a cursory ARV inspection. If so, obtain the Air Release Valve Report from the distribution system senior technician for preventive maintenance before leaving the yard. Notify dispatch of the job location and request dispatch to relay location and continued work status to the customer service department.

As required, set up "lane closed" signs, cones/barricades, construction signs, etc., as per the Department of Highways and Public Transportation Traffic Control Handbook, ensuring that the jobsite and area is safe for workers, pedestrians, and motorists. Contact the maintenance coordinator or district supervisor for assistance.

Inform the customer about what type of work you will be doing, and when you should be finished with the job.

Work Steps:

1. Check the area for any potential hazards.
2. Visually inspect the chamber lid for any possible damage and remove.
3. Before entering the chamber, follow the guidelines for entering confined spaces as described under Confined Space Entry—Activity.
4. Prepare chamber to facilitate the detailed ARV inspection.
5. Add or modify all air vent location information as needed.
6. Complete the ARV observations section of the worksheet, including the valve type, make, model, diameter, installation angle, connection type, status, soil and rock level, and polywrap condition.
7. Collect a soil sample adjacent to the water main for evaluation.
8. Perform all gate valve observations and document data.
9. Perform tapping saddle strap measurements if present to assess corrosion advancement and document all data collected.
10. Photograph discrepancies or deficiencies for future assessments.

Work Completion:

1. Replace the lid on the valve box, manhole, or vault.
2. Clean up the work area.
3. Inform the customer of what was done, and if additional crews will be back to complete the job. File photographs with report.

The distribution system senior technician for preventive maintenance completes the Air Release Valve Inspection Worksheet (or appropriate worksheet, work ticket, etc.), verifying the existing data and logging the work performed operation code, etc.

If deficiencies are found with the valve, valve box, chamber, etc., that cannot be repaired, fill out a service request form and turn it in to the senior distribution system manager.

Questions to Check Progress

1. Does the utility have implemented and documented policies and procedures to exercise all distribution and transmission valves in an exercising program that is comprehensive, integrated, and preventive on a systemwide basis?

2. Does the utility have implemented and documented goals to exercise the number of valves in the distribution system on an annual basis and identify the deficient valves and address these deficiencies in a timely manner?

3. Does the utility have policies and procedures to initiate corrective action if the annual goals of valve exercising or repair or replacement in a prescribed timely manner are not met?

4. Does the utility identify all the critical valves in the distribution system and regularly exercise them on a frequency greater than other valves?

4.2.6	**Fire Hydrant Maintenance and Testing**
4.2.6.1	Maintenance and testing. The utility shall have a hydrant maintenance and fire flow testing program. Testing and maintenance shall comply with the requirements of AWWA Manual M17, *Installation, Field Testing, and Maintenance of Fire Hydrants*. This program shall include as a minimum the following elements:

1. A goal for the number of hydrants to be inspected and tested based on a percentage of the total hydrants in the system.

2. Procedures for opening and closing hydrants to minimize potential damage to the distribution system.

3. Fire flow-testing requirements.

Rationale

Providing adequate fire protection is a major element of any water utility's function. One major component of this element is the fire hydrant and its maintenance. Periodic testing and maintenance is required to ensure proper operation when needed during emergency conditions.

All hydrants should be inspected regularly, at least once a year, to ensure their satisfactory operation, and those needing repairs should be properly tagged and the fire department informed. It is essential that all repairs are documented, which means that each hydrant should have its own unique identifier for record-keeping purposes.

Fire flow tests are conducted to determine pressure and flow-producing capabilities at any location within the distribution system. The primary function of fire flow tests is to determine how much water is available for fighting fires, but the tests also serve as a means of determining the general condition of the distribution system. Flow tests can help detect closed valves as well as heavily tuberculated mains and are used extensively by insurance underwriters as a factor in setting rates for insurance premiums.

Example of Methods or Procedures

Fire flow testing follows the initial procedures of a flushing program, with the proper notifications and work preparation. The following procedures are for the testing steps.

1. Screw a pressure gauge setup on one 2½-in. outlet of the upstream static hydrant (hydrant closest to a transmission main or a larger-diameter main).

2. Fully open the static hydrant, bleeding the air through the petcock valve of the setup.

3. Observe the static pressure at the static hydrant, recording the pressure two or three times at 30-second intervals to ensure average pressures are obtained.

4. Record all information relevant to the test on the appropriate worksheet.

5. Place appropriate signs ("water on road," "flooded area," etc.) on either side of the flow hydrant site(s).

6. Contact dispatch to inform them of the work being done and what hydrants are affected.

7. Screw either a diffuser on each 2½-in. outlet of the flow hydrant, or use one 4½-in. diffuser. If the static pressure is remaining stable, the person at the static hydrant will call or signal the personnel at the flow hydrant(s) to begin the test.

8. Open the flow hydrant and flushing valve fully, and bleed off the air from the pitot by opening the petcock valve.

9. Measure the flow from the diffuser at the flow hydrant(s). Make sure the water is flowing unobstructed into a catch basin, ditches, etc., and remove obstacles as necessary. NOTE: During freezing temperatures, ensure all flowing water remains out of the roadway.

10. When flowing the 2½-in. nozzle(s), use the information in Table 4.2.4-1 to convert psi to gpm (gallons per minute). When flowing the 4½-in. nozzle, use Table 4.2.4-2 to convert psi to gpm. Tables are located at the end of Sec. 4.2.4.

11. When the residual pressure reading has stabilized, the person at the static hydrant will call the personnel at the flow hydrant(s) to ensure flow hydrant(s) are fully open, and to tell them to take the pitot reading. DO NOT allow the residual pressure to fall below 20 psi (138 kPa) during the test, per State Primary Drinking Water Regulations, R.61-58.5(D).

12. Confirm that an adequate pressure drop has been obtained at the static hydrant (a minimum pressure drop of 10 psi [69 kPa] is needed to accurately calculate the Q20-value).

13. In the event the pressure drops were too small using one flow hydrant, be prepared to flow an additional hydrant. This situation may occur on large distribution mains and transmission mains.

14. Close the hydrants slowly to minimize any water hammering.

Questions to Check Progress

1. Does the utility have implemented and documented policies and procedures to exercise all fire hydrants in an exercising program that is comprehensive, integrated, and preventive on a systemwide basis?

2. Does the utility have implemented and documented goals to exercise the number of fire hydrants in the distribution system on an annual basis and identify the deficient fire hydrants and address these deficiencies in a timely manner?

3. Does the utility have policies and procedures to initiate corrective action if the annual goals of fire hydrant exercising or repair or replacement in a prescribed timely manner are not met?

4. Does the utility identify all the critical fire hydrants in the distribution system and regularly exercise them on a frequency greater than other fire hydrants?

5. Does the utility collect routine fire flow data?

6. Does the utility closely work with the local fire departments and inform them of the status of all fire hydrants in terms of operational status functionality and locations?

7. Does the utility have implemented and written procedures to address all above scenarios that are based on AWWA Manual M17?

4.2.7   **Materials in Contact With Potable Water**

4.2.7.1   Approved coatings or linings. The utility shall use AWWA Standards, NSF Standard 61, or other appropriate standards to specify approved coatings or linings for distribution system components that come into contact with potable water.

Rationale

The NSF Water Treatment and Distribution Systems Program is responsible for the certification of drinking water treatment chemicals and drinking water system components to ensure that such products do not contribute contaminants to drinking water that could cause adverse health effects. NSF/ANSI Standard 61: *Drinking Water System Components—Health Effects* is the nationally recognized health effects standard for all devices, components, and materials that contact drinking water. By following this standard, a distribution system will help to ensure the health and safety of its customers.

Example of Methods or Procedures

The following statement should be included in the minimum standards for the design and construction of the water utility:

Materials

General:

a. All chemicals/products added to public water supply must be third-party certified as meeting the specifications of NSF/ANSI Standard 60.

b. All materials/products that contact potable water must be third-party certified as meeting the specifications of NSF/ANSI Standard 61.

Questions to Check Progress

1. Does the utility only use materials, casings, and linings that are in contact with potable water as per AWWA standards, NSF standards, or as approved by the local health and regulatory agencies?

2. Does the utility have documented and implemented policies and procedures to ensure that all materials in contact with potable water safeguards the water quality in the distribution system?

4.2.8	**Metering**
4.2.8.1	Metering requirements. Utilities shall meter the volume of water entering the distribution system and accumulate historical data related to the volume of water used throughout the year to determine daily peak flows and maximum-day peak flows.
4.2.8.2	Metering devices. All metering devices shall meet the requirements of AWWA or other applicable standards.
4.2.8.3	Testing. To ensure meter accuracy, the utility shall have a goal to test or replace meters at the frequencies recommended in AWWA Manual M6, *Water Meters—Selection, Installation, Testing, and Maintenance.*
4.2.8.4	Repair and replacement programs. The utility shall have a program to replace or repair defective meters. The program shall include the necessary records to verify conformance with the guidelines as defined in AWWA Manual M6.

Rationale

Water meters are mechanical devices; therefore, loss of accuracy is dependent on use rather than length of time in service. Meters should be tested to ensure accurate registration for two main reasons. A meter for a large-volume customer that is underregistering by only a small percentage can result in a significant loss of revenue for the utility due to the high volume of water passing through the inaccurate meter. On the other hand, if the meter is overregistering even by a small percentage, then a large overcharge to the customer may result. Meters are tested for accuracy to ensure fair and equitable billing. Meters that test out of the accuracy range must be calibrated or repaired and retested.

Older meters for which repair parts are no longer available should be removed from the system and replaced with newer meters. Once these mechanical devices have reached their life expectancy, they should be removed from the system and systematically replaced to ensure accurate metering and accounting for the water being distributed.

Example of Methods or Procedures
Example 1

The following are steps for testing water meters ⅝ to 2 in. in size to ensure the meters are registering accurately. Test results are input into a database and periodically analyzed to ensure optimal meter life and consumption registration.

1. Gather all necessary equipment. All meters to be tested must be properly cleaned and pass preliminary inspections for broken, scratched, or foggy glass; damaged body; etc., before testing.

2. Inspect the test bench to see if it is in proper working order, i.e., water flows through the pipes and the hydraulic clamps work. The drain valves for the test tanks should be able to open and close easily. Make sure the pressure valves on the test bench are in the open position before you place the meters on the bench.

3. Ensure inlet and outlet pressure gauges have valid calibration stickers on them.

4. Check inlet pressure gauges. If the pressure is below 50 psi (345 kPa), monitor pressure. If it remains below 50, reschedule test for a later time. Low pressure will result in reduced flows.

Meter Testing (⅝ to 1 in.):

1. Remove the orange plugs from the inlet and outlet sides of the meter.

2. Drain the appropriate tank being used by opening the drain valve located at the base of each tank.

3. Hold the flow indicator with one hand and move the riser pipe over the test tank being used. Make sure the test tank drain valve is open.

4. Turn the hydraulic clamp valve to the closed position.

5. Completely open the inlet main water valve located next to the hydraulic clamp valve. Water will flow out of the bench drain valves, located between each meter station, into the test bench basin.

6. Make sure the test tank being used is empty. Now, close the test tank drain valve.

7. On the appropriate test sheet, write down the meter number, manufacturer, current date, etc.

8. Clean the glass with a wet paper towel. If testing one meter, run water until the sweep hand is on zero.

Low-Flow Test:

1. Hold the flow indicator still, move the riser pipe over the small calibrated tank, and close its drain valve.

2. Open the outlet main water valve until the flow indicator is reading steady at 1 gpm.

3. Close the valve quickly when the water level is even with the black line of the test tank for the 1-m^3 mark.

4. Take down the readings on each meter and find the accuracy (meter reading divided by tank volume).

5. Open the drain valve on the tank to allow the water to drain.

Medium-Flow Test:

1. With the riser pipe still over the small tank, close the drain valve.

2. Open the outlet main water valve so that the flow indicator is reading steady at 2 gpm.

3. Close the valve quickly when the water level is even with the black line for the 1-m^3 mark.

4. Take down the readings on each meter and find the accuracy.

5. Open the drain valve on the tank and drain all the water.

High-Flow Test:

1. Hold the flow indicator still and move the riser pipe over the large test tank (10 ft^3).

2. Close the drain valve on the large tank.

3. Completely open the outlet main water valve to the highest flow possible (approximately 20 gpm).

4. Close the outlet main water valve quickly when the water level is even with the black line for the 10-ft^3 mark.

5. Take down the readings on each meter and find the accuracy.

6. Open the drain valve on the tank to allow the water to drain.

Passing/Failing of Meters: Only meters that have an accuracy between 98 percent and 102 percent will pass. All others will be recycled or repaired.

Work Completion:

1. Close the inlet main water valve behind the hydraulic valve.

2. Open the pressure valves near the meters.

3. Open the hydraulic valve. The meters will separate from the clamps and the water in the meters will pour into the test bench basin. If you do not open the pressure valves before this step, then you will become very wet, as the pressure will not be released.

4. Take the meters and pour the water from each side.

5. Spray the inlet and outlet sides of each meter with chlorine solution, and spray the orange plugs before inserting them into the meters.

6. Insert the orange plugs into the meters.

7. Box meters and take to inventory.

8. Clean, reorganize, and put up all tools used.

Example 2

These are procedures to verify that small water meters are in proper working order with no defective parts or leaks.

1. Visually examine the box and log its condition, box style, and whether it's in the grass, dirt, or pavement.
2. Clear all overgrown grass, debris, or other obstructions from around meter box and, using a screwdriver, remove the box lid.
3. If the box is full of water or the meter is submerged, use the hand pump to remove all water and log that the box is not draining properly.
4. Remove all dirt and debris from inside the meter box so that the entire meter and both inlet and outlet valves are visible.
5. When an old style heavy/light meter box is encountered, locate the curb stop (if exists) in front of the meter box.
6. Remove the curb well lid and make sure the curb well is straight and clear of dirt and debris, and the curb stop T-head is accessible with the ¾-in. meter key.
7. Test the inlet and the outlet valve(s) (if exists) with the ¾-in. meter key to ensure they are functioning properly.
8. Inspect the meter and check for leaks at the meter flanges. Log any deficiencies found on the work ticket or appropriate form.

If leaks are found at the meter flange, then:

1. Put on rubber gloves to avoid possible electrical shock.
2. Tighten spinner or flange bolts and nuts.

If leak persists:

1. Turn off inlet and outlet (if exists) valve(s); notify the customer prior to shutting off the water.
2. Loosen the meter spinner or flange bolts and nuts.
3. Remove meter.
4. Remove the inlet and outlet gaskets.
5. Clean the inlet and outlet sides of the setter (use a scraper if necessary).
6. Install new gaskets.
7. Replace meter.
8. Open spigot(s) on customer's property.
9. Reactivate service.
   - Check the register lid to ensure it is intact and log the meter number located on top of the lid (when in the closed position).

- Examine the meter register and check to make sure the glass is not scratched, cracked, or fogged, and all dials and numbers are clear and legible.
- If register is in poor condition:
    (1) Put on rubber gloves to avoid possible electrical shock.
    (2) Using a screwdriver, remove pin on the side of the register.
    (3) Turn register in counterclockwise direction ¼ turn.
    (4) Remove register.
    (5) Clean top of meter under register with wire brush.
    (6) Log meter reading of old and new register on work ticket.
    (7) Place new register on meter and turn clockwise.
    (8) Install new pin.
    (9) Make sure register is operating; turn on outside spigot if necessary.

Questions to Check Progress

1. Does the utility have a distribution system that is capable of delivering maximum-day and fire flows for individual public fire requirements?

2. Does the utility have documented and implemented policies and procedures to ensure that any alteration or expansion of the distribution system will provide adequate flow to meet maximum-day and fire flow for individual public fire requirements?

3. Does the utility have fire flows to meet the Insurance Service Office (ISO) requirements of the community as per the ISO ratings?

4. Does the utility have a goal to improve that ISO rating of the distribution part of the community?

4.2.9	**Flow**
4.2.9.1	Flow requirements. The system shall be capable of delivering the maximum-day demand and fire flow for individual and public fire requirements.

Rationale

There is no legal requirement that a governing body must size the distribution system to provide fire protection; however, most communities chose to. For large systems, providing fire protection has a marginal effect on the design of the distribution system. But for small systems, providing fire flow requirements in addition to the maximum-day demand can mean a significant increase in storage capacity and minimum pipe diameters.

To ensure that proper flow requirements are being met, the maximum-day demand and the fire flow requirements need to be determined prior to designing the distribution system. There are several methods established for determining fire flow requirements. It is important that the system be able to provide the proper flow requirements in a worst-case scenario, so that public heath and safety is continuously maintained.

Example of Methods or Procedures

Example 1

The following is an excerpt from a utility's minimum design standards. This is an example of how the fire flow requirements will be taken into account to properly size the mains of a distribution system.

C. Sizing of Lines

1. Size piping based on either $\frac{1}{5}$ the instantaneous maximum flow plus fire flow or maximum instantaneous demand, whichever is greater. When fire protection is to be provided, system design should be such that fire flows and facilities are in accordance with the requirements of the utility and the state Insurance Service Office (ISO).

2. Minimum design fire flow shall be 1,000 gpm (3,785 L/min) with a minimum residual pressure of 20 psi (138 kPa).

3. Design for 2.5 fps flushing velocity in accordance with State and Federal regulations.

4. All water mains, including those not designed to provide fire protection, shall be sized using a hydraulic analysis based on flow demands and pressure requirements.

5. The developer's design engineer is to determine available static and residual pressures at the delivery point for the water to a new development. The data is to be obtained under the direction of an engineer who is registered in the state.

6. Use Hazen and Williams design coefficient, C = 120.

7. The maximum instantaneous demand is to be calculated using the Tables 4.2.9-1, 4.2.9-2, 4.2.9-3, as published in the *Community Water System Source Book* by Joseph S. Ameen.

Table 4.2.9-1   Maximum Instantaneous Flows for Residential Areas

Number of Residences Served	Flow per Residence, gpm
1 (First)	15.0
2–10*	5.0
11–20†	4.0
21–30	3.8
31–40	3.4
41–50	3.2
51–60	2.7
61–70	2.5
71–80	2.2
81–90	2.1
91–100	2.0
101–125	1.8
126–150	1.6
151–175	1.4
176–200	1.3
201–300	1.2
301–400	1.0
401–500	0.8
501–750	0.7
751–1,000	0.5

*Second, third, etc., through tenth residence served.
†Eleventh, twelfth, etc., through twentieth residence served.

Table 4.2.9-2   Maximum Instantaneous Flows for Commercial Areas

Type of Business	gpm on Basis Shown
Barber Shop	3.0 gpm per chair
Beauty Shop	3.0 gpm per chair
Dentist Office	4.0 gpm per chair
Department Store*	1.0–3.0 gpm per employee
Drug Store	5.0 gpm
With Fountain Service	Add 6.0 gpm per fountain area
Serving Meals	Add 2.0 gpm per seat
Industrial Plants†	4.0 gpm plus 1.0 gpm per employee
Laundry	30.0 gpm per 1,000 pounds clothes
Launderette	8.0 gpm per unit
Meat Market, Supermarket	60 gpm per 2,500 sq. ft. floor area
Motel, Hotel	4.0 gpm per unit
Office Building	0.5 gpm per 100 sq. ft. floor area or 2.0 gpm per employee
Physicians Office	3.0 gpm per examining room
Restaurant	2.0 gpm per seat
Single Service	6.0 to 20.0 gpm total
Restaurant Drive-In	2.0 to 7.0 gpm total
Service Station	10.0 gpm per wash rack
Theater	0.2 gpm per seat
Theater Drive-In	0.2 gpm per car space
Other Establishments‡	Estimate at 4.0 gpm each

*Including customer service.
†Not including process water.
‡Non-water-using establishments.

Table 4.2.9-3 Maximum Instantaneous Flows for Institutions

Type of Institution	Basis of Flow, gpm
Boarding Schools, Colleges	2.0 gpm per student
Churches	0.4 gpm per member
Clubs: Country, Civic	0.6 gpm per member
Hospitals	4.0 gpm per bed
Nursing Homes	2.0 gpm per bed
Prisons	3.0 gpm per inmate
Rooming House	Same as Residential*

Schools: Day, Elementary, Junior, Senior High	
Number of Students	Gpm per Student
0–50	2.00
100	1.90
200	1.88
300	1.80
400	1.72
500	1.64
600	1.56
700	1.44
800	1.38
900	1.32
1,000	1.20
1,200	1.04
1,400	0.86
1,600	0.70
1,800	0.54
2,000	0.40

*Each unit of an apartment building should be considered as an individual residence.

Example 2

The following is an excerpt from a utility's minimum design standards. This is an example of how the fire flow requirements will be taken into account to properly size hydrant leads.

A. General

3. Fire hydrant leads to be a minimum of 6 in. in diameter. Larger size mains will be required as necessary to allow the withdrawal of the required fire flow while maintaining the minimum residual pressure. A hydrant control valve shall be installed on all hydrant leads. The utility shall have final determination of main sizing.

Questions to Check Progress

1. Does the utility have a distribution system that is capable of delivering maximum-day and fire flows for individual public fire requirements?

2. Does the utility have documented and implemented policies and procedures to ensure that any alteration or expansion of the distribution system will provide adequate flow to meet maximum-day and fire flow for individual public fire requirements?

3. Does the utility have fire flows to meet the ISO requirements of the community as per the ISO ratings?

4. Does the utility have a goal to improve that ISO rating of the distribution part of the community?

4.2.10   External Corrosion

4.2.10.1   Leaks/breaks. The utility shall have a standardized system for recording and reporting pipeline leak or break information. At a minimum, the data collected on a leak or break report shall include pipe location, pipe material, pipe size, apparent type of leak or break, visual assessment of surrounding soil type (sand, clay, etc.), pipe's depth, and best assessment of saturation conditions of the soil prior to break or proximity to water table.

4.2.10.2   Monitoring program. Utilities shall have an external corrosion monitoring program. The program shall include surveys of pipeline route before construction, pipeline and metallic tanks not under cathodic protection, and pipeline and metallic tanks under cathodic protection. Corrosion surveys shall include potential measurements, line current measurements, soil resistivity, and soil chemical analysis. This data may be used to evaluate an infrastructure improvement program.

Rationale

Deterioration of pipelines, valves, pumps, and associated equipment due to external corrosion is a great concern for water distribution systems. Techniques are available to eliminate or significantly reduce this type of corrosion, which is generally defined as an electrochemical reaction that deteriorates a metal or an alloy. External corrosion of greatest concern for water utilities is that of dissimilar metals in contact with each other in a common media, great variances in soil in contact with metal, aggressive soil conditions, atmospheric corrosion, chemically contaminated soil, and microbiologically induced corrosion.

Corrosion is said to cost water and wastewater utilities around $36 billion annually. In addition to the financial impact of having to provide for the repair, replacement, labor, and equipment that is required due to external corrosion, the cost to public safety and health is even more important to consider. The health of consumers is at risk every time water pressure is inverted due to necessary repairs to corroded infrastructure. Also, in the same vein, the risk of a disaster increases with the loss of fire flows due to the effects of corrosion.

Certain tests and observations can help to indicate where corrosion will be a problem in a new or existing area of the system. When conditions indicate that corrosion will be likely, steps can be taken to reduce the severity or eliminate it completely. There are a variety of options available for corrosion control. On principle, corrosion can be prevented by eliminating one of the four elements required to support it—an anode, a cathode, an electrolyte, and a return current path. Coatings, sacrificial anodes, and polyethylene encasement are among the most popular and cost-effective options.

Example of Methods or Procedures

Generally put, potential is the force available to drive an electrical current through a circuit. In reference to a water distribution system, the circuit is between the pipeline and the surrounding soil. Surveying for potentials throughout a distribution system will help identify at-risk areas and allow for the proper prevention systems to be put into place. The following is an example of a potential surveying procedure.

Title: Potential Surveying

Purpose: To provide guidance, steps, and instructions for surveying the water distribution system in search of potential pipeline corrosion areas. In potential surveys, measurements are made of the electrical potential (voltage) between the buried pipeline and its environment. This is done with a suitable voltmeter having the negative terminal connected to the pipeline and the positive terminal connected to a copper sulfate reference electrode (half-cell) that contacts the environment. This activity identifies the proper procedures to follow for performing potential surveys within a specified area, and includes the procedures and directions to follow when using specific potential detection equipment.

Responsibility and Authority: For optimum performance, the surveyor must be proficient in the use of potential survey detection methods and equipment. The surveyor should take precautions when working in high-traffic areas, and must protect the equipment from tampering, damage, or theft.

Work Steps: Obtain copies of the grids that are scheduled for surveying. Resources used for scheduling are as follows:

- Corrosion survey grids
- Age-of-mains grids
- Main break report
- Potential survey spreadsheet

Potential Surveys Priorities and Scheduling: Priorities within each area are

1. Corrosive/Aggressive soils (known).
2. Cathodic protection data (revealing potential corrosive soils).
3. Age of water mains.
4. Landfill areas.
5. Size of water mains.

Potential Surveys Procedures: Begin by reviewing the grids and valve cards to verify main location and confirm the type of pipe material being surveyed. Measure 600 ft (183 m) of main and mark in 50-ft (15-m) increments.

Connect the wire reel to a pertinent structure (e.g., hydrant, meter, or valve) using the clamp on the wheel. Connect the test conductor lead for the wire wheel to the positive test lead of the voltmeter. Remove the cap from the bottom of the copper sulfate electrode (half-cell) and turn on the voltmeter, setting it to ohms with a range of 40 mΩ.

Beginning at station 0+00, plug the test conductor lead into the wheel. Place the copper sulfate electrode (half-cell) over the main adjacent to the wheel wire on the ground with the terminal side up. Connect the negative wire of the voltmeter to the terminal of the (half-cell) recording millivolt reading. Unplug the test conductor lead from the wheel.

Continue unreeling the wire to the next station number, placing the wire over the main as close as possible, and repeat the reading process. After recording the final station number reading, turn off the voltmeter, respool the wire onto the wheel, and disconnect the clamp from the structure.

Work Completion: Clean up work area, and collect tools and equipment used. Be sure all meter and valve box lids are replaced and secured. Upon returning to the operations center, input data into the potential survey spreadsheet and print out the data and graph. Highlight the corrosion survey grid map(s) using the following color codes:

- Blue—designated for all negative readings <0.00, which indicates a non-corrosive environment
- Red—designated for all positive readings >0.00, which indicates a corrosive environment

Questions to Check Progress

1. Does the utility have a comprehensive system and documented procedure to collect data related to distribution system failure, including the failure location; pipe materials; pipe size; apparent type of leak or break; visual assessment of surrounding soil type, i.e., sand, clay, etc.; pipe depth; and best assessment of saturation condition of the soil prior to break or proximity to water table?

2. Does the utility have comprehensive policies and written procedures to monitor external corrosion of the water distribution system that, at least, includes potential surveys?

3. If the utility has metallic tanks in the distribution system, is there a comprehensive cathodic protection program to control the corrosion?

4. Does the utility use corrosion monitoring data to provide effective protection of the existing and newly installed water mains and other infrastructure in the distribution system?

4.2.11	**Design Review for Water Quality**
4.2.11.1	Policies and procedures. Utilities shall have a formal, standardized design procedure that provides for comprehensive review of all construction projects to reduce the potential for water quality degradation during and following installation of the project.
4.2.11.2	Records. Utilities shall prepare as-built drawings of all installed facilities and shall maintain records associated with inspection, design, and construction of all new and retrofitted facilities.

Rationale

Water quality can be affected by a very large number of variables. Many of these variables, which can negatively or positively influence water quality in the distribution system, are determined before the system is even constructed or operated. If the design of the system takes into account the effects on water quality, many potential quality issues down the road can be altogether avoided. Effects on water quality should be examined when decisions on items such as pipe material, pipe size, dead ends, blow-offs, system geometry, etc., are being made.

A key part of being able to review design plans for water quality is to know what is already in place and how those facilities are affecting the water quality. In order to do that, accurate and thorough records need to be kept. As-built drawings, showing all details of a new, rehabilitated, or replaced section of the distribution system, are the best source for important information that will later be needed when operating the system.

Example of Methods or Procedures

The following is an example of a procedure to ensure that all water distribution project plans are reviewed by appropriate personnel to ensure that the company's standards are going to be met.

Title: New System Plan Review

Purpose: To provide guidance, steps, and instructions in the evaluation of proposed new water system's design, layout, installation, etc.

Scope: This activity identifies the proper procedures for reviewing preconstruction plans forwarded to the water distribution department from the design and construction department. The review process provides feedback and input from the water distribution department, ensuring that a quality water system is constructed and received.

The scope of this project is to identify possible conflicts in relation to engineering design, proximity of existing or proposed utilities, land encroachment/easements, and environmental aspects prior to construction.

Work Preparation: This activity is performed when an engineer or developer submits to the design and construction department a set of plans for review to ensure that the project meets company specifications. Design and construction department projects are also reviewed.

The development coordinator in the design and construction department tracks the projects and issues to the technical section supervisor a set of plans for the proposed new system with a cover sheet, including a requested completion date.

Work Steps: The senior distribution technician submits the plans to the distribution technician—technical support, who reviews the plans as follows.

1. Type of tie-ins or connections to the existing water mains:
   a. Type of tap and detail supplies.
   b. Type of tee and detail showing tie-in.
   c. Proper existing main or valve locations.
2. Proper valve location and spacing:
   a. ±500-ft (152-m) intervals.
   b. ±25 customers shut off at any one time.
   c. Review industrial layout.
3. Proper hydrant location and spacing:
   a. ±1,000-ft (305-m) intervals.
   b. 500-ft (152-m) "hose-laying length" to any one property line.
   c. 10-ft (3-m) separation between sewer laterals and hydrants.
   d. Review industrial layout.
4. Blow-off location and type:
   a. All dead ends must have either a hydrant or blow-off.
   b. Kupferle Main-guard blow-off to be used (or equal).
   c. Detail provided.
5. Proximity to other utilities.
6. Easements required, including size.
7. Location of meters, vaults, etc.:
   a. Location suitability.
   b. Access for meter department.
   c. Proximity to traffic.
8. Proximity of mains, valves, hydrants, etc., as compared to sidewalks, streets, curbing, etc., Look for conflicts with future repairs.
9. Proximity and inclusion of 3-in. PVC service-line carrier pipes:
   a. Coverage of all lots.
   b. Conflicts with sewer laterals (should be on opposite property lines from sewer laterals).

    c. Location in relation to property lines (should be in-line with the opposite property lines).

10. Check profiles for possible conflicts. Profiles delineate a cross-sectional view of utilities from finished grade. NOTE: There should exist an 18-in. (46-cm) horizontal separation between the bottom of water mains and the top of sewer mains, with water over the sewer.

11. Inclusion of construction details:
    a. Hydrant
    b. Valve and valve box
    c. Fitting taps
    d. Tie-ins
    e. Thrust block
    f. Easements
    g. Pipe bedding
    h. Service lines
    i. Service carrier pipes
    j. Street patching
    k. Blow-off

Work Completion:

1. Make review notes on supplied plans in red.
2. Make notes on cover sheet referencing:
    a. Location of discrepancy
    b. Description of discrepancy
    c. Suggested solution
3. Submit plans and cover sheet to the technical section supervisor.

Questions to Check Progress

1. Does the utility have comprehensive policies and implemented and written procedures for designing new water mains to reduce the potential for water quality degradation during the installation and operation?

2. Has the utility developed and utilized comprehensive guidelines for design and installation of existing and new water distribution systems based on the best management practices and AWWA standards meeting or exceeding local codes and regulatory requirements?

3. Does the utility have policies and procedures regarding the preparation of as-built drawings of all installed facilities and do they ensure that all the existing and new water distribution as-built drawings and other records are accurately kept and readily available?

4. Does the utility have documented and implemented policies and procedures to design, inspect, and commission all retrofit and new facilities in the distribution system?

4.2.12 Energy Management

4.2.12.1 Energy management program. The utility shall have a program to review and optimize electrical energy usage. The program shall have the following elements:

1. A review of energy usage, identification of energy use trends and cost or usage tracking versus time.

2. Consideration of energy costs in its evaluation of new distribution system facilities.

Rationale

Over the past few years, awareness of greenhouse gases, climate change, global warming, and carbon footprints has increased. Water professionals are now analyzing whether or not their actions in providing safe drinking water inadvertently affect other aspects of the environment. Not only is environmental stewardship the right thing to do, but energy is a significant portion of a water utility's operating expense. As a rule, electricity consumption is the second highest cost for water distribution companies, behind labor. The price of electricity is also rising at a far higher rate than labor costs, making it likely that it will become the dominant cost in production and distribution of water in the near future.

Energy management is at the heart of efforts across the entire sector to ensure that utility operations are sustainable in the future. More and more utilities are realizing that a systematic approach for managing the full range of energy challenges they face is the best way to ensure that these issues are addressed on an ongoing basis in order to reduce climate impacts, save money, and remain sustainable. A review of a facility's energy performance may also identify other areas for operational improvements and cost savings, such as labor, chemicals, maintenance, and disposal costs. A thorough assessment of a facility's energy performance may alert managers to these issues. For example, an unexplained increase in energy consumption may be indicative of equipment failure, an obstruction, or some other problem.

Example of Methods or Procedures

The following report shows how one utility attempted to reduce energy usage at one of their pump stations during a rebuild by adding a metallic polymer to the impeller. The coating was advertised to reduce power consumption and improve efficiency.

Title: Evaluation of Pump Rebuild

Background. In early 2005, the treatment plant staff determined that the impeller and volute on Pump 13 high service pump (manufactured by Goulds) needed to be rebuilt because of severe wear. The plant staff hired ITT Industries Pro Service Center (through Tencarva Machinery) to perform the inspection and to rebuild the impeller and volute. ITT Industries was chosen because Goulds Pump is a division of this company. The rebuild contract included the application of the Belzona coating to the impeller and volute and the installation of new wear rings, bearings, seals, and gaskets.

ITT recommended the application of a Belzona 1341 supermetalglide coating to the impeller and volute of Pump 13. The 1341 coating is a metallic polymer that is NSF approved. It is applied in two coats after the metal surfaces have been blasted. The main purpose of the coating is to minimize abrasion wear. In addition, the coating is advertised to "reduce power consumption and improve efficiency by improving hydrodynamic performance." Treatment plant staff agreed to the Belzona application as a way to reduce abrasion since the Pump 13 impeller was displaying a severe wear condition.

Plant staff decided to evaluate the effect of the rebuild work on the pump's energy use. However, power consumption information on Pump 13 could not be obtained before the pump was shipped out for rebuild because the power meter was not functioning and had to be replaced. As a result, there were no data available to perform a pre- and post-rebuild comparison of power use.

In 2006, the impeller and volute on Pump 11 high service pump was sent out for a rebuild. Plant staff decided to have the same Belzona coating applied to Pump 11. Plant staff were able to take seventy-eight readings from November 2006 to January 2007 on the Pump 11 before it was shipped off. The readings consisted of pressure and power data. Flow information was collected, but these data could not be used because it was discovered that the magnetic flowmeters could not provide accurate flow data. After the pump was rebuilt, plant staff were able to collect sixty-eight readings. These readings were taken in May and June 2007. In addition, SCADA staff set up trend charts for kilowatt data.

Basis of Comparison. As previously stated, the plant staff decided to monitor Pump 11 performance before and after the rebuild to see if power consumption by the pump was affected by the rebuild work. The initial idea for comparing "before" and "after" conditions was to use the TDH (total dynamic head) versus flow curve. However, it became evident that flow readings would be a weak link because high service flow was being measured with Marsh-McBirney insertion magnetic meters when the data collection effort began. Over the course of three months, a statistical control project proved that the meter's readings were erroneous because the settings for the pipe diameter had to be radically adjusted from the actual diameter so that flow data would be read close to what is typically expected. In addition, flowmeter readings from the mag meters did not correctly react to varying pressure conditions. For example, if the TDH increased, the flow readings might increase when logically they should always decrease. The statistical control project eventually proved the validity of existing Venturi flowmeters that are mounted on the same high service lines, but this did not happen until after the "before rebuild" data readings were taken on Pump 11.

Because flow data were not reliable, it became obvious that the study would not be able to quantify the change in the pump efficiency as a result of the rebuild because the pump power formula requires kilowatt input, flow, and TDH information in order to calculate efficiency. Since reliable flow data were not available prior to the rebuild, plant staff decided to use pressure and kilowatt data to determine how the rebuild work affected pump performance. Since Pump 11 already had gauges mounted on the suction and discharge sides, gauge readings were taken before and after the rebuild. Since the gauges were not connected to SCADA, instantaneous pressure readings had to be recorded. Instantaneous readings for kilowatts were also obtained from the Pump 11 power meter at the same time. The gauge readings were corrected to the centerline of the pump so that an overall TDH could be calculated for each reading.

Together the TDH and the kilowatt reading provided a satisfactory means of comparing before and after conditions in terms of power consumption. In theory, the power consumption should always be the same at a given TDH, and this scenario should be repeatable over and over as long as the instruments are in proper working condition. The only thing that would change this scenario would be a change in the flow characteristics of the pump. The wear ring replacement would have an impact on flow performance, but there was no way to quantify this without an accurate flow performance test before and after the rebuild. Since there was no way to measure flow accurately, plant staff decided to compare the trend in kilowatt consumption at various TDH points to quantify the rebuild impact.

Results. The TDH–kilowatt data were plotted on an Excel graph showing data for the pre-rebuild pump in the November 2006–January 2007 data set and the post-rebuild pump in the May–June 2007 data set (Figure 4.2.12). From the graph, it is easy to see that the TDH–kilowatt points for the pre-rebuild data set are above those of the post-rebuild data set. The data demonstrates that Pump 11 was clearly using more power prior to the rebuild of the pump. SCADA data for the same time periods as the data sets confirms the Excel chart results.

The graph is a good visual indicator of how power consumption has been reduced, but it does not provide a numeric value for the power reduction. A value was determined by comparing the average kilowatts consumed at TDH points that were common to both data sets. A list of those points with the average kilowatts consumed is shown in Table 4.2.12. In addition, the difference between the kilowatts for each data set is calculated and averaged to provide a value for power reduction.

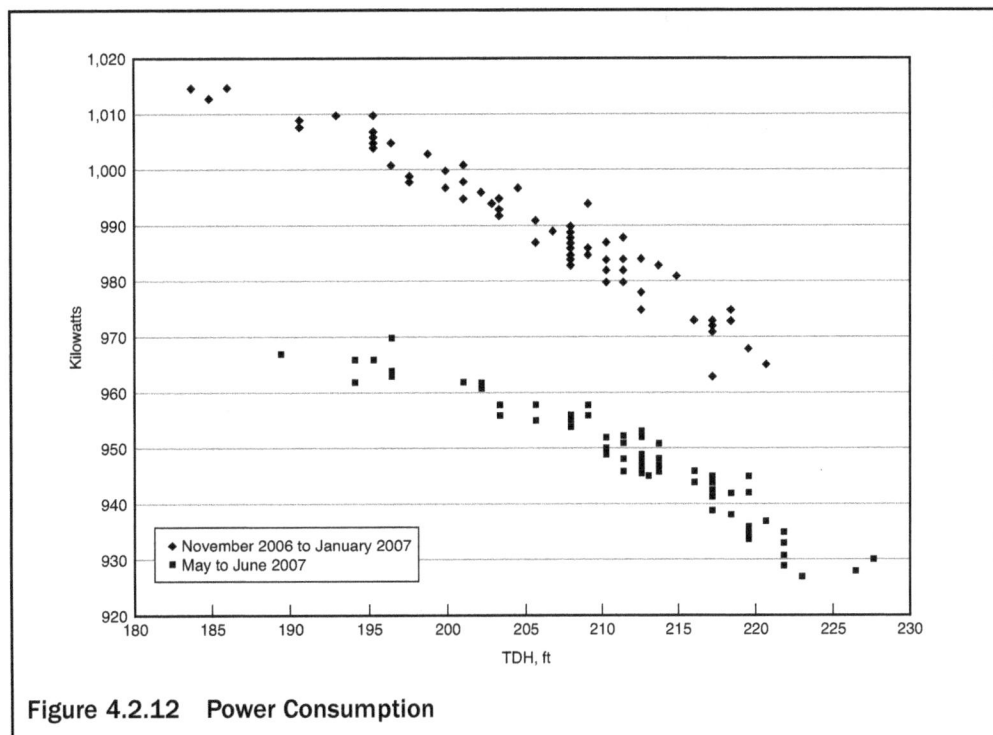

Figure 4.2.12   Power Consumption

Table 4.2.12   Reduction in Power

November 2006 to January 2007		May 2007 to June 2007	
TDH, ft	Kilowatts	Kilowatts	Difference
221	965	937	28
220	968	938	30
218	974	939	35
217	970	942	28
216	973	945	28
214	983	948	35
213	979	949	30
211	984	949	35
210	983	950	33
209	988	957	31
208	986	955	31
206	989	957	32
203	993	957	36
202	996	962	34
201	997	962	35
196	1,003	967	36
195	1,006	966	40
		Average	33

Table 4.2.12 shows that the average reduction in power has been 33 kilowatts. Based on this average, the cost savings per year can be estimated using a $0.05 per kilowatt-hour unit price. The run time for Pump 11 has averaged 2,624 hours per year from 2004 to 2006. The estimated cost savings is $4,330 per year.

Questions to Check Progress

1. Does the utility have comprehensive policies and procedures to monitor the energy use in the O&M of the distribution system?
2. Does the utility conduct energy audits on a regular basis for the O&M of the distribution system, and have goals to reduce the energy use?
3. Does the utility use energy conservation policies and technology in proactive ways in design and operation of new facilities in the distribution system?

Sec. 4.3   Facility Operations and Maintenance

4.3.1   **Treated Water Storage Facilities**

4.3.1.1   Storage capacity. The utility shall establish minimum operating levels in storage facilities based on pressure in the distribution system, fire flow requirements, emergency storage requirements, and other site-specific conditions.

4.3.1.2   Operating procedures. The utility shall have written operating procedures, which address water level fluctuations in the storage facilities and turnover rates. The utility shall have a goal to reduce water age in the finished water storage facility. The utility shall have a written policy in place that establishes the target turnover rate along with minimum and maximum operating levels.

4.3.1.3   Inspection. The utility shall have a written inspection program outlining frequency, procedures and maintenance of records. The inspection program shall include such features as routine (daily/weekly); periodic (monthly/quarterly); and comprehensive (3–5 years) inspections.

4.3.1.4   Maintenance. The utility shall have a maintenance program that includes periodic cleaning and refurbishing of facilities, as required. Cleaning of covered storage shall be based on internal inspection conducted at a minimum of every 5 years and for uncovered reservoirs, at least annually. The utility shall perform a full internal and external inspection according to AWWA Manual M42, *Steel Water-Storage Tanks*. The utility shall conduct an external visual inspection of the storage facility at least seasonally to assess and repair environmental damage and verify the integrity of vents and screens. The inspection shall include an assessment of the physical security of the facility. Maintenance activity, such as coating or painting, shall be based on ANSI/AWWA Standards D102 and D103.

4.3.1.5   Disinfection. The storage facility shall be disinfected according to ANSI/AWWA C652 if it is drained for the inspection. Disposal of heavily chlorinated water shall be done in accordance with local, state, and federal regulations. If divers or remote operational vehicles (ROVs) are used, the divers and equipment shall be disinfected according to ANSI/AWWA C652.

4.3.1.6   Additional requirements. All treated water storage facilities shall be covered and protected from contamination or shall incorporate additional treatment of the water as it leaves the reservoir.

Rationale

The purposes for finished water storage in the distribution system can vary per system, for example, ensuring reliability of supply, maintaining pressure, equalizing pumping and treatment rates, reducing the size of transmission mains, and improving operational flexibility and efficiency. If properly designed and operated, finished water storage tanks can accomplish all of the above. However, one major point of concern is the susceptibility to water quality changes and degradation if not operated and maintained appropriately.

There are a wide number of factors that influence the design, operation, and maintenance of finished water storage facilities. Of greatest concern should be system capacity and water quality. Operation of storage facilities should be such that fire flows and the maximum-day demand are available throughout the distribution system at any time, while minimizing negative impacts on water quality.

Inspection of finished water storage facilities will depend on the type of facility, susceptibility to vandalism, age, condition, history, and other types of criteria. However, there are three types of inspections that should be conducted on a regular basis. Routine inspections can be conducted from the ground on a weekly basis, with other routine tasks. Periodic inspections are slightly more rigorous and may involve climbing the tank every quarter or so. Comprehensive inspections are major undertakings and involve inspection of the interior of the storage facility, which may require it to be drained and taken off line. Inspections of this nature will require a team of experts and ideally should occur every three to five years. Many times this is also when a utility should plan for the cleaning and, if necessary, recoating of a finished water facility. However, based on periodic inspections and water quality, cleaning may be required more frequently.

Example of Methods or Procedures
Example 1

All utilities should establish sufficient finished water storage capacity based on system-specific criteria. Two main factors are state regulatory requirements for sizing and storage capacity and ISO fire protection standards.

One utility designs to levels beyond ISO standards by ensuring that storage facilities can supply fire flows of 4,000 gpm (15,142 L/min) for four hours, in addition to meeting state requirements of having half the daily demand available in storage.

Example 2

The following is an example of work steps for cleaning and inspecting clearwells, typically done on a three-year cycle.

1. Pump liquid level down as low as practicable.
2. Coordinate with operations on the valve alignment for draining the clearwell.
3. Drain or pump out remaining volume. Dechlorinate as needed.
4. Use clean equipment inside the storage facility.
5. Clean floor using appropriate tools, hoses, pressure washer, and/or pumps to remove sediment.
6. Clean walls and baffles (as applicable) by pressure washing to remove film.
7. Sediment or removed materials may be pumped out into an appropriate drain. Solids should be contained. Water may be allowed to discharge to storm drains or overland drainage.
8. Flush inlet/outlet pipe(s) to remove sediment.
9. Pressure wash interior of concrete ground storage tanks.
10. Concrete structure inspection shall be conducted by a registered engineer. Perform or contract any necessary repairs as per applicable standards.
11. Metal structure inspection shall be conducted by a registered engineer. Perform or contract any necessary repairs as per applicable standards.
12. Disinfect per AWWA standard and conduct bacteriological sampling and testing.

Upon satisfactory disinfection and bacteriological test results, the water may be released to the distribution system if it is of acceptable aesthetic quality.

Questions to Check Progress

1. Does the utility have written policies and procedures to operate treated water systems to meet pressure requirements, fire flow requirements, energy storage requirements, or other site-specific conditions?
2. Does the utility have written policies and procedures to maintain adequate volume and water quality, which may include the reduction of the age of water and increase the turnover rate in the treated water facility?
3. Does the utility have a comprehensive PM program with written policies and procedures to periodically inspect and clean the reservoir at least once every five years, internal and external corrosion monitoring

and control, and at least annual inspections based on AWWA Manual M42, which includes vent and screen inspections, periodic cleaning, and painting based on ANSI/AWWA standards?

4. Does the utility disinfect the treated water storage reservoir after cleaning and disposal of chlorinated water as per ANSI/AWWA C652?

5. Does the utility have documented policies and procedures to protect the water storage reservoir from contamination?

6. Does the utility have implemented policies and procedures to secure all water storage facilities and monitor water quality routinely?

4.3.2   Pump Station Operations and Maintenance

4.3.2.1   Operating procedures. The utility shall establish written operating and maintenance procedures that document all functioning of each pumping facility (including emergency power generating equipment). Operating logs shall record operational conditions, such as inlet pressure, discharge pressure, individual pump run times, flow rate, and other operational variables.

4.3.2.2   Maintenance program. The utility shall have written maintenance procedures for each pumping station (including emergency power generating equipment) describing frequency, procedures, and maintenance of records. Information shall include basic manufacturer operating requirements. Records shall document all inspections and any service performed.

Rationale

Booster pump stations usually consist of a series of in-line pumps used to boost pressures or flows at specific points throughout the distribution system. Not all systems use booster pumps. However, if booster pumps are required by a system to maintain system pressures and/or fire protection, then operational logs and records are key tools to keep the pumps and the overall system running efficiently.

Sudden and/or unexplained changes in operational conditions can be a red flag indicating major issues within the distribution system. For example, if a booster pump is cycling on more frequently, then this may indicate a large main break downstream of the station.

Well-kept maintenance records will give operators a good indication of actions that may need to be taken to keep the system running at optimal conditions. For example, if a certain pump is requiring additional corrective maintenance at more and more frequent intervals, this may indicate another problem that has not been addressed or the need for replacement.

Example of Methods or Procedures

The following is an example of the procedures for monthly inspection of pump stations.

Title: Pump Station Inspection

Read these instructions before starting work. If the inspector has any questions, he/she should consult the supervisor for clarification.

Safety Requirements: Follow all requirements of the safety manual and the applicable safety program or procedure. This may include, but is not limited to, lock-out tag-out, confined space entry, hazard communications (material safety data sheets), process safety management, excavation and shoring, respiratory protection, electrical safety, and lab safety.

Use personal protective equipment predicated by the work environment and material handling requirements. This may include, but is not limited to, eye protection, face protection, hard hat, foot protection, welder's equipment, life preservers, respirators, gas masks, self-contained breathing apparatus, gas detection equipment, seat belts, hearing protection, and traffic vest.

Generator Checks:

1. Check engine fluids before starting generators. Run and record readings.
2. Log engine hours in the logbook and record proper information on this sheet.
3. Check the coolant level in the radiator by removing the radiator cap and checking the level. The level should be no more than ½-in. (1.27 cm) from the bottom of the fill neck.
4. Check the sight glass on the expansion tank for the cooling water level, located on top of radiator.
5. Remove the oil dipstick from the engine and ensure that the oil level is between the high and low marks on the dipstick. If below the low mark, then refill the oil level to the full mark on the dipstick.
6. Check the gas fuel system for leaks by visually inspecting the lines. If a strong odor is noticed, then there may be a gas leak on the fuel line. Isolate and repair as necessary.
7. Visually inspect the engine for oil or water leaks.

8. Check the batteries for their water levels. Refill the batteries with distilled water as needed.

9. Check the engine for proper operation. Run generator for six hours. Check the engine for leaks while running at temperature. Record information required.

10. After running the generator, put engine in a cool-down cycle. After completing this cycle, place generator switch back into the automatic mode.

Questions to Check Progress

1. Does the utility have implemented and documented policies and procedures for the O&M of the pumping stations as recommended by the manufacturer?

2. Does the utility have implemented and documented procedures to monitor the operation of the pumping facilities, which at least includes inlet pressure, discharge pressure, individual pump run times, flow rate, and other operational variables?

3. Does the utility have implemented and documented procedures for maintenance and corrective maintenance programs for each pumping facility and emergency power generation equipment, describing frequency, procedures, and maintenance records?

4. Does the utility have documented emergency procedures to be implemented during a pumping station failure?

4.3.3	**Pipeline Rehabilitation and Replacement**
4.3.3.1	Rehabilitation and replacement program. The utility shall have a program for

Rehabilitation and replacement program. The utility shall have a program for evaluating and upgrading existing portions of the distribution system as required. The program shall include provisions for maintaining records to access the physical condition of the pipes. Records shall include the following information:

1. Current system maps.

2. Maintenance records for all leaks and breaks. The repair type, pipe condition, and the joint type are noted.

3. Distribution system data (fire flow tests, C-factor tests, pressure gauges, and other pertinent data).

4. Surrounding environmental information, i.e., data on soil types, corrosion potential, and the location of hazardous material sites.

5. Proper separation between potable water lines and other pipelines carrying nonpotable water or other hazardous material shall be maintained in accordance with the authority having jurisdiction.

6. The number and type of water main breaks. The utility shall establish an annual goal in terms of breaks/100 mi/yr (breaks/100 km/yr) of distribution pipe.

Rationale

Water utilities are among the most capital-intensive enterprises in existence. The centerpiece of their mission is to build and sustain the extensive infrastructure systems that provide services necessary to life and public health. Of late, utilities have been in an expansion mode, with little consideration of sustaining the existing infrastructure. Now the picture is changing and utilities have been slow to make renewal a major issue. In many cases they have only limited knowledge of the condition or even the extent of their assets. As renewal becomes a major issue, they are unable to make a convincing case for the magnitude of reinvestment necessary now to prevent falling even farther behind as they head into the future.

Many pieces of data are needed to make smart rehabilitation and replacement decisions. The very first is a complete asset inventory. Systems maps and records need to be kept accurate and as up-to-date as possible. There are many factors that may indicate the need for replacement or renewal of areas of the distribution system. All these factors come together to give a manager or engineer a realistic picture of the condition of the system.

Example of Methods or Procedures

The following gives an example of a project prioritization strategy for the replacement and/or rehabilitation of system water mains using all the data that should be collected to get a clear picture of the mains' condition.

Title: Criteria for Renewal/Replacement Prioritization

The following 10 parameters are considered when appraising each individual project for rehabilitation. Each section is weighted on a 10-point scale; the higher priority is given a higher rating, so the worst-case scenario would achieve a total score of 100 points. Any project that scores more than 70 points will be considered, and the higher the point value assigned to the project, the higher the priority level.

I. Main Break Density: The first parameter to be evaluated is the number of main breaks that were reported in the previous five-year period. By taking the total number of breaks during the past five years and dividing by five, an average number of breaks per year can be employed as the first indicator. These numbers can presently be recovered via the departmental "Main Break" database and, as the computerized maintenance management system is populated, it can also be retrieved via the report writing software currently used by the Water Distribution Department, "Crystal Reports."

Main Break Density	
0	0 Main Breaks/mile—5 yr. Average
1	
2	1 Main Break/mile—5 yr. Average
3	
4	2 Main Breaks/mile—5 yr. Average
5	
6	3 Main Breaks/mile—5 yr. Average
7	
8	4 Main Breaks/mile—5 yr. Average
9	
10	≥5 Main Breaks/mile—5 yr. Average

II. Age of Main: This factor takes into consideration the age of the water main or its appurtenances. The older the section of main in question, the higher the score. The typical lifespan of a water main depends on the quality of the material used in manufacturing. The industry standard for a water main is approximately 100 years; therefore, the table below incorporates a progressive scale based on age.

Age of Main	
0	≤1 year
1	≤10 years
2	≤20 years
3	≤30 years
4	≤40 years
5	≤50 years
6	≤60 years
7	≤70 years
8	≤80 years
9	≤90 years
10	>90 years

III. Loss of Effective Pipe Size: This criterion is based on internal pipe condition due to tuberculation as derived from coupon samples and/or pipe samples. These samples represent the condition of a water main throughout its length. Pipe tuberculation shrinks the ID (internal diameter) of the water main, which diminishes the carrying capacity and fire flow available via the main in question.

Loss of Effective Pipe Size	
0	0% loss of effective pipe size
1	
2	
3	
4	
5	25% loss of effective pipe size
6	
7	
8	
9	
10	50% loss of effective pipe size

IV. Corrosion Status–Soil Condition: The fourth consideration is a subjective criterion that deals with the condition of the surrounding environment. This determination can be assigned based on data such as known potential surveys, soil resistivity–moisture content, existing soil conditions, the proximity of adjacent utilities, and known cathodic protection devices installed.

Corrosion Status–Soil Condition	
0	Good condition
1	
2	
3	
4	
5	Poor condition
6	
7	
8	
9	
10	Severe condition

V. Water Quality Concerns: The fifth decisive factor relates to the amount of water quality concerns that have been documented at each location. Biofilm growth and tuberculation are key sources of unfavorable water calls from customers, and unlined water mains are the typical source of this type of complaint. Data related to water quality concerns are currently available, and historical data can be obtained dating back to approximately 1995.

Water Quality Concerns	
0 Good	0 Customer concerns/3-year period
1	
2	
3	
4	
5 Poor	≤5 Customer concerns/3-year period
6	
7	
8	
9	
10 Severe	≥10 Customer concerns/3-year period

VI. Hydraulics—Pressure/Flow/C-Value: The following section pertains to the hydraulic statistics that are obtained on the pipe section in question. This criterion is similar to the "loss of effective pipe size" in that as the ID is reduced, so is the carrying capacity of that line. The reduced capacity directly relates to the ability that the water main has to provide adequate fire protection.

Hydraulics—Pressure/ Flow/C-Value	
0	Pressure ≥ 70 psi, Flow ≥ 1,200 gpm, C-Value ≥100
1	
2	
3	
4	
5	Pressure ≥ 60 psi, Flow = 1,000–1,200 gpm, C-Value = 50–80
6	
7	
8	
9	
10	Pressure ≥ 50 psi, Flow ≥ 1,000 gpm, C-Value ≥ 50

VII. Appurtenance Condition Code: This criterion is based on the known operation codes for the attached appurtenances. You will notice that these existing codes have been subdivided into three categories: good, moderate, and poor, which somewhat eliminates the subjectivity from the evaluation.

Hydrant Operation Code Table	Valve Operation Code Table
1. Good (Good)	1. Good (Good)
2. Fair (Good)	2. Fair (Good)
3. Poor (Moderate)	3. Poor (Moderate)
4. Frozen (Poor)	4. Frozen (Poor)
5. Wrung (Poor)	5. Wrung (Poor)
6. Rounded Op. Nut (Poor)	6. Packing Leak (Moderate)
7. Leak at Op. Nut (Moderate)	7. Unable to Locate (Moderate)
8. Leak at Break Flange (Moderate)	8. Rounded Op. Nut (Moderate)
9. Leak at Main Valve/Boot (Poor)	9. Op. Nut Missing (Moderate)
10. Nozzle Damaged (Moderate)	10. Bent Stem (Poor)
11. Hydrant Too Low (Moderate)	11. Missing HD Nut (Moderate)
12. Leak at Nozzle (Moderate)	12. Overseats (Moderate)
13. Hydrant Too Low (Moderate)	13. Guts Removed (Poor)
14. Rotated/Leaning (Moderate)	14. Closed (Moderate)
15. Other (Poor)	15. DO NOT OPERATE
16. Obstructed (Moderate)	16. Mult. Op. Codes (Poor)
17. Inoperable (Poor)	17. Leaking (Poor)
18. Cap(s) Missing (Good)	18. Other (Poor)
19. Multiple Op. Codes (Poor)	19. Unable to Op. (Poor)
20. PM Needed (Moderate)	20. Needs Raised (Poor)
21. Schedule(d) to Replace (Poor)	21. Needs Lowered (Poor)
22. Flushing (Good)	
23. Needs Tag (Good)	
24. Needs Raised (Moderate)	
25. Needs Lowered (Moderate)	

Appurtenance Condition Code	
0	Good Operation Code
1	
2	
3	
4	
5	Moderate Operation Code
6	
7	
8	
9	
10	Poor Operation Code

VIII. Pipe Material: This criterion is based on the existing pipe material used. This factor is also divided into three groups: good, moderate, and poor. By including this section in the evaluation process, the replacement of material that does not meet the utility's current specifications can be prioritized in the rehabilitation/replacement process.

Pipe Material	
0 Good	Ductile iron, PVC
1	
2	
3	
4	
5 Moderate	Cast iron
6	
7	
8	
9	
10 Poor	Asbestos cement, galvanized, conduit

IX. Zone Code: This segment of the evaluation process is categorized into three zones, depending on the type of dwelling supplied. Higher point values and priority will be given to those critical sections of the distribution system that supply high-risk and/or heavy industrial customers; however, each project must be evaluated on the specific conditions that pertain to the surrounding area since rural and residential zones are critical to our customers as well.

Zone Code	
0	
1 Residential	Rural
2	Residential
3	Dense residential
4 Institutional	Church
5	Municipal government
6	School/College
7 Commercial/Industrial	Commercial
8	Industrial
9	Medical
10	Hospital

X. Maintenance Cost: The final criterion is based on the amount of maintenance performed on the main during the past year. This section is subjective at this point due to the amount of job cost reports that may be required to properly evaluate each project. Once the maintenance management system is fully implemented, with its accounting capabilities, the cost per job should become more easily retrievable and quantitative criteria can be implemented.

Maintenance Cost— Low/Medium/High	
0 Low	
1	
2	
3	
4	
5 Medium	
6	
7	
8	
9	
10 High	

Questions to Check Progress

1. Does the utility have rehabilitation and replacement programs based on certain percentages to rehabilitate and replace the distribution system?

2. Do these programs prioritize projects based on the best current rehabilitation and replacement technologies and consider such factors as current

system maps, maintenance records of leaks and breaks, distribution system data, and site conditions?

3. Does the utility have a five-year capital program and adequate resources provided for its implementation for the rehabilitation and replacement of the distribution system and associated facilities?

4.3.4	**Disinfection of New or Repaired Pipes**
4.3.4.1	Disinfection of new or repaired pipes. All new and repaired pipe sections shall be protected from contamination and adequately disinfected. Pipe and pipe sections shall be disinfected in accordance with the requirements of ANSI/AWWA C651 Disinfecting Water Mains.
4.3.4.2	Bacteriological testing. Bacteriological testing shall be completed in accordance with ANSI/AWWA C651 Disinfecting Water Mains.
4.3.4.3	Disposal of chlorinated water. Disposal of heavily chlorinated water shall be done in accordance with local, state, and federal regulations.

Rationale

Water mains cannot reasonably be kept sterile during installation. In fact, mains may be left outdoors for months before they are installed, or may be flooded with water of poor quality during construction. Only disinfection after the pipe is installed can ensure clean pipes to deliver safe drinking water. The same goes for main repairs when pressure losses occur.

Highly chlorinated water is the most common means used to disinfect new or repaired water mains, in which case this toxic by-product must be removed from the system through flushing. Ideally the water should be discharged to a sanitary sewer collection system, as long as the system can handle the discharge. Direct discharge of highly chlorinated water to a body of water or storm sewer system may be in violation of the Clean Water Act. Highly chlorinated water can easily be neutralized prior to discharge into the environment.

Example of Methods or Procedures

Disinfection consists of four tasks:

1. Preventing contamination of the new pipe during shipping, storage, and construction.

2. Flushing the water main to remove loose debris and dirt that may have entered the water main during construction. NOTE: Do not flush if hypochlorite tablets have been placed.

3. Chlorination of the water main to destroy pathogenic microorganisms. There are three accepted methods for adding the chlorine chemical to new water mains (ANSI/AWWA C651):

   • Calcium hypochlorite tablets placed during construction—this method provides an average dose of 25 mg/L

   • Continuously fed sodium or calcium hypochlorite solution—this method provides a minimum chlorine residual of 10 mg/L for 24 hours

   • A slug of high concentration chlorine solution—this method gives a high concentration of chlorine, 100 mg/L, for a contact time of greater than three hours

4. Bacteriological testing of the disinfected water to ensure that the microbiological water quality is adequate.

Chlorination procedures:

1. Choose the chlorination method best suited for the installation.

2. Ensure all boundary valves are closed. Use a water distribution system drawing to highlight all valves and pipes involved in the area to be disinfected.

3. Tag all boundary or isolated valves: "DO NOT OPERATE."

4. Set up the chlorination equipment in such a way that the feed-point is not more than 10 ft (3.5 m) downstream from the beginning of the new water main.

5. DO NOT use fire hydrants for chemical feed. The high concentration chlorine solution will damage the hydrant.

6. Notify dispatch and open one boundary valve and the discharge valve permitting water from the distribution system or other approved source to flow through the new water main at a constant, measured rate. Use a pitot gauge, a bucket of known volume, and a stopwatch to calculate the rate of discharge.

7. Adjust the chemical feed pump rate to produce a chlorine dose of 25 mg/L free chlorine.

8. Frequently monitor the discharge location for chlorine residual using an approved field test kit.

9. Once the 25-mg/L residual has been achieved, stop the discharge and chlorine feed. Retain the chlorinated water in the test section for 24 hours or more.

10. Operate all valves and hydrants in the test section to ensure they are disinfected.

11. After 24 hours, check the free chlorine residual for a residual of 10 mg/L or greater. If the residual is less than 10 mg/L, flush and rechlorinate.

Questions to Check Progress

1. Does the utility have documented and implemented policies and procedures for the disinfection of both existing mains when they are exposed to the outside environment and new mains when they are installed as per the requirements of ANSI/AWWA C651?

2. Does the utility have documented and implemented procedures to properly dispose of chlorinated water as per the local regulatory requirements?

3. Does the utility have documented and implemented procedures for the bacteriological sampling and testing of existing and new mains as required by ANSI/AWWA C651 and *Standards Methods*?

4. Does the utility have documented and implemented procedures for the disinfection of the pipes and materials during the repair of the distribution system?

SECTION 5: VERIFICATION

Sec. 5.1   Documentation Required

The utility shall define critical activities and create written procedures for each. The utility shall have a training component for personnel (Sec. 5.2.2). The utility shall maintain an adequate record-keeping system so that compliance with this standard can be measured.

5.1.1   General

The documentation shall include

  a. Documented statements of a quality policy and quality objectives.

  b. Standard operating procedures.

  c. Documented procedures required by this standard.

  d. Documents needed by the utility to ensure the effective planning, operation, and control of its processes.

  e. Records required by this standard.

NOTE: Where the term *documented procedure* appears within this standard, it means that the procedure is established, documented, implemented, and maintained.

Rationale

Documentation is a useful way to manage and control processes. Writing standard procedures and work instructions will make sure that everyone is performing the process the same way, helping to ensure quality and competency. Any documentation for the standard O&M of the distribution system should include best management practices, technology, efficiency, and effectiveness to ensure that the system is performing at high standards. Documentation should be approved, reviewed, and revised if necessary on a regular schedule and be available at all points of use within the organization. Documentation that is standardized in format will be easier to maintain.

Documented goals can provide a clear road map to take an organization from where they are to where they need to be. If goals are not documented, it is easy to lose sight of the big picture and flounder on unimportant tasks.

Example of Methods or Procedures

Example 1

The following is an example of a written quality policy and objective for a water distribution system.

Title: Water Distribution Department Goals

- The ultimate goal of the Water Distribution System is to ensure a continuous supply of treated potable water to our customers. This is accomplished at the lowest possible cost to the consumer without compromising water quality or customer service in the process. Of equal importance is the need to maintain a safe working environment for our associates, and those exposed to our operations. As a "front-line" representative of the utility, it is our responsibility to reflect these goals on a continuous basis.

- Our business and the prime purpose of our existence are to provide service to our customers. Customers expect and deserve value for their money, and judge us by the level of satisfaction they receive. Customers are the focal point of all endeavors and will be treated with courtesy, concern, and competence.

- All of our business is accomplished through people. Associates are our most valued assets, and we are committed to their personal development, well-being, and self-realization. We believe that frank and fair associate relations, recognition of achievement, open communications, training and development, and participative decision making are essential to creating progressive, motivated, and dedicated associates.

- We will strive for excellence by promoting the unity of authority, responsibility, and accountability for each job and place the emphasis on high standards and achievement at all levels and in all areas.

- We will strive to develop strong leadership, characterized by a propensity for action, leading by example, a high profile within the company's community, and commitment to the philosophy and values.

- We are committed to the prevention of pollution, protection of the environment, and continual environmental improvement for present and future generations.

- We will strive to be the best-in-class Water Distribution System in the country.

Example 2

Table 5.1.1 is an example of a table of contents for Standard Operating Procedures (SOPs).

Table 5.1.1 SOP Manual Table of Contents

Activity/Section No.	Activity/Section Description
	Foreword
Section 1.1	Vision & Mission Statements
Section 1.2	Water Distribution Department Goals
Section 1.3	EMS Policy Statement—Uncontrolled Copy
Section 1.4	Company History
Section 1.5	Description of the Water System
Section 1.6	Organizational Chart
Section 1.7	Flow Charts:
	WD Department Task Flow Charts
	Maintenance Task Flow Chart
	District Engineering Flow Chart
	Inventory Flow Chart
	Meter Technology Flow Chart
	Administration Task Flow Chart
Section 2.1	Activity Scope & Definitions Page
2.1.1	Main-break Control
2.1.2	Service Leak Control
2.1.3	Hydrant Leak Control
2.1.4	Valve Leak Control
2.1.5	Settlement Investigation
2.1.6	Temporary Water Hose Hookup
2.1.7	Temporary Water Supply Tank (Buffalo)
2.1.8	Solar Arrow Board

Example 3

Figure 5.1.1 is an example of a standard form on which to create standard operation procedures. Using a standard form will help to ensure that all key points of the procedures are covered.

Title: Standard Operating Procedures Instruction Form

Purpose/Scope: This form is used to prepare Standard Operating Procedures (SOPs) for use by associates who perform operational control and maintenance management tasks. The department head (or designee) is responsible for the approval, revision, and issuance of SOPs and ensuring that associates have the necessary training to perform the job.

Instructions: Complete each section below prior to placing new or modified monitoring or control equipment in service. The department head (or designee) shall ensure the maintenance documents are available and controlled at all appropriate locations. NOTE: This form may be replicated on a computer or duplicated

Prepared by:	Approved by/Date:	
Title:		
1. Purpose:		
2. Scope:		
3. Responsibility and Authority:		
4. Work Preparation		
4.1 Labor:	4.2 Equipment:	4.3 Materials:
5. Work Step(s):		
6. Related Documents:		

Figure 5.1.1   Standard Operating Procedures Instruction Form

on a photocopier. The computer copy must look similar to this document and contain the same information.

Questions to Check Progress

1. Does the utility have documented and critical activities to be conducted for the proper O&M of the distribution system?

2. Does the utility have documented and implemented procedures for all the activities performed for the O&M of the distribution system by integrating the best management practices, technology, and effective and efficient resources?

3. Are the standard operating procedures for the critical activities as identified earlier readily and easily available?

4. Does the utility have these procedures established, documented, implemented, and maintained in a standard format with proper versions to ensure that everybody is using the most recent version of the standard operating procedure?

5.1.2   **Examples of Documentation**

Documentation shall be sufficient to support the requirements in Sec. 4, including

- Regulatory compliance records
- Monitoring plan and test results
- Sample locations—frequency, low-residual sites, long travel time sites
- Disinfectant residual results—maximum, minimum, average statistics
- Ammonia, HPC results
- Booster disinfection goals, residual results
- Disinfection by-product results—maximum, minimum, averages
- Corrosion monitoring results
- Color, taste, and odor results
- Customer water quality inquiries and responses
- Backflow prevention program and testing records
- Flushing program procedures and results
- Pipe materials specifications
- Storage tank detention time, cleaning records, treatment evidence
- Leak detection and water loss calculations
- Valve exercise goals and numbers
- Hydrant exercise goals and numbers
- Meter testing records
- Pressure records—maximum, minimum, averages
- Flow records—maximum, minimum, averages
- External corrosion—testing records
- Design review for water quality—construction project water quality checkoff
- Energy management program—strategy to optimize usage
- Pump station operation and maintenance plan—records
- Pipeline restoration and replacement program—actual miles vs. planned
- Pipeline disinfection records

Rationale

All of the above listed items are critical elements of this standard for operating and maintaining a water distribution system. A system cannot possibly ensure that they have implemented and are maintaining all of these standards appropriately without proper documentation. Without writing these standards into policies and procedures, a system cannot know that the standards will be continuously maintained or properly implemented.

Example of Methods or Procedures

The following is a sample of the table of contents for a monthly report that tracks important performance indicators for a water distribution system.

Title: Monthly Operating Report

1. Current Year's Strategic Plan
2. Administrative Section
   a. Key Performance Matrix
   b. Training Program—Section Summaries
   c. Certification Statistics
   d. Safety Program
   e. Injuries/Incidents/Violations/Inspections
   f. Department Highlights
   g. Recognition and Development Program for Department
   h. Crew Performance Indicator
   i. Productive Measurement Comparisons
   j. Preventive Maintenance vs. Corrective Maintenance
   k. Critical Business Area Monitoring
   l. Equipment Downtime and Usage Monitoring
3. Environmental Management System Section
   a. Computer Maintenance Management System/Geographic Information System (CMMS/GIS) Monitoring
   b. Regulatory Self-Assessment
4. Maintenance Section
   a. Main-Break Monitoring
   b. Main-Break Repair Tracking
   c. Anode Installation
   d. New Service Activity and Status and Statistics

  e. Damage to Other Utility Monitoring

  f. Damage to Our Utility Monitoring

5. Preventive Maintenance Section

  a. Water Quality Monitoring and Maps

  b. Department of Health and Environmental Control (DHEC) Monthly Sampling Requirement

  c. Unidirectional Flushing Program

  d. UDF Improvement Program

  e. Transmission Valve Exercising

  f. Air Release Valve Monitoring

  g. Hydrant Painting Program

  h. Unaccounted-for Water Monitoring Program

6. Technical Support Section

  a. Leak Survey Program

  b. Corrosion Control Program

  c. Hydrant/Valve Change Out Programs

  d. Utility Protection Monitoring

7. New Installations Section

  a. Main Installation/Replacement/Rehabilitation Program

  b. Asphalt/Concrete Patching Program

  c. Customer Patching Concerns

  d. Large Meter Installation Program

  e. Project Management Program

8. Meter Technology Section

  a. Large Meter PM Program

  b. Meter Testing Revenue Impact Report

  c. Large Meter Replacement Program

  d. Meter Vault Inspection Program

  e. New Service Average Installation Time

Question to Check Progress

Does the utility have comprehensive documents including operational, implementation, documentation, and effectiveness, which may include the folders:

- Regulatory compliance records
- Monitoring plan and test results

- Sample locations—frequency, low-residual sites, long travel time sites
- Disinfectant residual results—maximum, minimum, average statistics
- Ammonia, HPC results
- Booster disinfection goals, residual results
- DBP results—maximum, minimum, averages
- Corrosion monitoring results
- Color, taste, and odor results
- Customer water quality inquiries and responses
- Backflow prevention program and testing records
- Flushing program procedures and results
- Pipe materials specifications
- Storage tank detention time, cleaning records, treatment evidence
- Leak detection and water loss calculations
- Valve exercise goals and numbers
- Hydrant exercise goals and numbers
- Meter testing records
- Pressure records—maximum, minimum, averages
- Flow records—maximum, minimum, averages
- External corrosion—testing records
- Design review for water quality—construction project water quality checkoff
- Energy management program—strategy to optimize usage
- Pump station O&M plan—records
- Pipeline restoration and replacement program—actual miles vs. planned
- Pipeline disinfection records

5.1.3   Control of Documents

Documents required for this standard shall be controlled. Records are a special type of document and shall be controlled according to the requirements given in 5.1.4.

Rationale

It is just as important to maintain control over the distribution, updating, and storage of documents as it is to create them in the first place. This is to make sure workers receive and use the proper and the latest procedures, drawings, references, related material, etc. Effective document control will result in reduced costs by helping to eliminate wasted effort and material due to the use of improper documents. It will also ultimately result in increased efficiency and high levels of performance.

Example of Methods or Procedures

This example provides guidelines, steps, and instructions for updating, adding, deleting, handling, and storing water distribution department documents and documents of external origin, such as forms, written SOPs, equipment O&M manuals, etc.

Title: External Origin Documentation

This activity's performance criteria are based on the state Public Records Act, the state's Department of Archives & History Local Government Records Manual, and the company's Environmental Management Procedure. All records created or received during the course of business are considered public records.

Responsibility and Authority: When documentation changes occur, the distribution system support technician will issue new updates as follows to each associate on the controlled distribution lists, and will retrieve obsolete activities, documents, etc., from the associates. All associates are responsible for the proper creation, use, maintenance, retention, preservation, and disposal of public records per the state Public Records Act. Also, the electronic forms files and hard copies will be changed if applicable.

If a document of external origin is considered necessary for operation, then it is maintained in a locked cabinet and it is listed on our file index, which is contained within our master list. When updated versions of these documents are received, we will replace the physical document we are controlling and update the file index.

Work Steps: Procedures, processes, and work instructions should be defined, appropriately documented, and updated as necessary. SOPs and the Emergency Preparedness Manual will be reviewed periodically by the department's supervisory staff.

Form Modification: The distribution system support technician receives a modified original or originates changes to an existing document. A document authorization form will be submitted with changes.

The document requiring modification, creation, or deletion will be drafted by the associate requesting the changes. NOTE: Any associate may request changes to a document.

All documents will be a standard format and contain the following where applicable:

- Corresponding Environmental Management Systems (EMS)/ISO procedures
- Activity number or space for activity number
- Effective date
- Prepared by
- Approved by
- Title of document
- Identification number
- Regulatory reference

The associate originating the changes will complete a document authorization form.

A copy will be submitted to the distribution system support technician and "controlled" or "uncontrolled" status will be determined.

The distribution system support technician will modify the draft version (if necessary) or return it to the originator for final modification (i.e., effective date, changes, etc.).

The distribution system support technician will issue a document number and complete a document authorization form (this signed form will stay with the original document), and it will be issued to the director of water distribution for approval.

If a new form, then the distribution system support technician will fill out a records series inventory form and send it the records retention specialist.

Questions to Check Progress

1. Does the utility have documented, implemented, and monitored procedures to ensure that all critical documents identified in the standards are controlled according to the requirement established?

2. Does the utility have readily available all the critical documents identified in this standard?

5.1.3   ## Control of Documents (continued)

A documented procedure shall be established to define the controls needed to

a. Approve documents for adequacy prior to issue.

b. Review and update as necessary and re-approve documents.

c. Ensure that changes and the current revision status of documents are identified.

d. Ensure that relevant versions of applicable documents are available at points of use.

e. Ensure that documents remain legible and readily identifiable.

f. Ensure that documents of external origin are identified and their distribution controlled.

g. Prevent the unintended use of obsolete documents and to apply suitable identification to them if they are retained for any purpose.

Rationale

For standard operation procedures and other types of important documentation to be effective for their intended purposes, certain controls need to be in place. The controls set in this standard will help to ensure that effort, time, and money are not wasted by workers following inappropriate procedures. Associates must have the most up-to-date versions available in order to perform their jobs to the best of their abilities. For that to happen, someone must periodically review, update, and approve any revision. Documents must also be clearly legible and easy to access and identify by all who need them.

Example of Methods or Procedures
Example 1

Figure 5.1.3-1 is an example of one way to control important documents.

Title: Controlled Document Authorization

Purpose/Scope: To ensure that all controlled documents have been reviewed, revised as necessary, and are approved for adequacy and release by the appropriate authority. This form is to be completed when generating controlled documents.

Instructions: Complete each section below. Upon approval, file the original of this form with the controlled hard copy of the document being issued for use.

Part A. Document Description

Document Title:

Purpose/Scope:

Originator of Document:

Document Number:

Total Pages to Follow:

Type of Document: (i.e., policy, form, SOP)

Revision Date:

Review Date:

Expiration Date (if applicable):

Part B. Review and Approval

My signature below indicates that the above-referenced document has been reviewed for design, content, appropriateness, and adequacy and that I am authorized to sign for the approval of this document. NOTE: Please sign below in BLUE INK.

Approved by:_____

	Signature	*Date*

Part C. Distribution
(This section is to be completed if additional controlled copies are distributed.
This section may be continued on a separate page if additional space is required.)

Copy Number (if applicable)	Assignment (i.e., associate name, Internet location, manual title)	Date

Figure 5.1.3-1   Controlled Document Authorization

NOTE: This form may be replicated on a computer or duplicated on a photocopier. The computer copy must look similar to this document and contain the same information.

Example 2

Figure 5.1.3-2 is an example of a master list showing all important documents that are required to be controlled.

Title: Master List of Controlled Documents

Purpose/Scope: To identify and record a master listing of controlled documents.

Instructions: An entry shall be made for each controlled document upon approval and updated thereafter as necessary. Additional entries may be made on a new form or on the Master List of Controlled Documents Continuation Sheet. NOTE: This form may be replicated on a computer or duplicated on a photocopier. The computer copy must look similar to this document and contain the same information.

Group(s): Water Distribution					
Prepared by/Date:			Approved by:		
			Effective Date:		
Reference/ I.D. Number	Title of Document	Type of Document	Approved by/ Date	Document Review and/or Exp. Date	Date Review Completed

Figure 5.1.3-2   Master List of Controlled Documents

Question to Check Progress

Does the utility have established, documented, and implemented appropriate procedures such as those documents listed in Sec. 5.1.3?

5.1.4   Control of Records

Records shall be established and maintained to provide evidence of conformity to requirements and of the effective operation of this standard. Records shall remain legible, readily identifiable, and retrievable. A documented procedure shall be established to define the controls needed for the identification, storage, protection, retrieval, retention time, and disposition of records.

Rationale

Just as documentation and control of documents is necessary for the successful O&M of a distribution system, records and control of records is necessary for the improvement of the O&M of a distribution system. In order to set goals and strive to achieve more, any organization has to determine a benchmark for key processes. Maintaining accurate records and controlling those records is the only way to do that.

Example of Methods or Procedures

The following is an example of a records control procedure.

Title: Document Control Procedure/Records Control Sections

Records Retrieval:

- Associates needing to review filed records will see a member of the administrative staff for the issuing of files. File cabinets will be kept locked for security.

- The administrative staff person will remove the required record/file, place a checkout card in its place, and issue the record/file. Personnel files will be issued only to supervisors. The checkout card must be completely filled out.

- The associate issued the file will return the file to the designated file return location, return it to the administrative associate that previously assisted them, or to another administrative assistant.

- If a file is not current and stored on-site, the distribution system support technician will retrieve the requested file from storage. When the file is returned, it will be returned to storage.

- If a file is in storage off-site, then the distribution system support technician will issue a request to the records management specialist for the file and return it to same when it has been reviewed.

Records Retention/Purging:

- The records retention specialist maintains the retention schedule.

- Emergency Manual SOPs and Distribution Manual SOPs will be reviewed according to the program review schedule.

- Filed records will be reviewed periodically for purging, and records with a permanent retention schedule will be packed for storing. Other records will be defaced and recycled.

Work Completion:

- The obsolete original document will be stamped "OBSOLETE" and put in the obsolete file in the EMS files or the historical files in the Central files.

- All other copies will be defaced and put in appropriate recycle bins.

- The distribution system support technician files the new controlled copy in the appropriate current controlled document file.

Questions to Check Progress

1. Does the utility have documented and implemented policies and procedures for adequate control of records?

2. Does the utility maintain adequate records to provide evidence of conformity and of the effective operation and implementation of the standard?

3. Does the utility have documented, established, and implemented controls needed for identification, storage, retrieval, retention, and disposal of records?

Sec. 5.2 Human Resources

5.2.1 General

Personnel performing work affecting distribution system operation shall be competent on the basis of appropriate education, training, skills, test requirements, and experience as required by the governing regulatory agency.

Rationale

For a water distribution system to operate successfully, the workers involved with the daily maintenance and operation must also be successful. To ensure that, requirements in education, training, and skills testing as set forth by local regulatory agencies need to be identified and listed for persons and/or positions operating and maintaining a distribution system.

Example of Methods or Procedures

The following is an example of a policy that requires associates to be at a certain level of operator certification, as determined by the job title and functions, in accordance with state law.

Title: Operator Certification Policy

Intent of the Program. The *Water Distribution Licensing Program* provides a method to ensure that newly hired associates achieve the proper level of certification, as required by the State and the utility. The program applies to all positions that require a license at the Water Distribution facilities (see attached job description table).

Program Guidelines. This program requires unlicensed associates hired into the Water Distribution Department to register within 15 days as an Operator Trainee with the State Department of Labor, Licensing and Regulation; Environmental Certification Board. The applications are available from the Water Distribution Department's Training Coordinator.

Registration fees, initial examination fee *(for each certification level)* and annual license renewal fees are paid by the utility. Cost for any repeated examinations are borne by the associate. Associates are expected to pass the examinations within the timeline specified by the program (see the attached job description table).

This document shall apply to all current associates from the day of this program's inception. All current associates in the Water Distribution Department whose positions require licensing will be encouraged to participate and may be required to adhere to this program. If an associate changes position within his or her licensing period, the licensing period may be allowed to start over for the new position. **Failure to pass the certification examination within the time period specified by the program may result in termination or reassignment.**

Licensed associates hired at the Water Distribution Department will meet the following criteria: **If the associate has a "D," "C," or "B" License when hired,** they must adhere to the time schedule specified by the minimum requirements of their job description (see the attached job description table [Table 5.2.1]). **Failure to pass the certification examination within the time period specified by the program may result in termination.**

Program Procedures. The following steps will be followed with the *Water Distribution Licensing Program.* New hires should apply for a Trainee Permit as a Water Distribution Operator within 15 days of their first day of employment with the utility. The utility will cover this cost. The employee should use the department's address when applying to ensure renewal fees requests come to the department for prompt payment.

Job Description Table. Every Associate whose position requires a license must adhere to the following schedule [see Table 5.2.1]. Failure to meet any of these deadlines may result in termination or reassignment.

Memorandum of Understanding. New hires will sign a "Memorandum of Understanding" at the time of department orientation. See the following example [Figure 5.2.1].

Training and Examinations. The Associate may make application for the "D" Level examination at the same time he/she applies for operator trainee registration. Examination can be scheduled daily and are scheduled by the Water Distribution Department's Distribution System Support Technician. Examinations are given twice a day at various locations throughout the state. The Company will cover this cost. The exam results are provided at the time of the exam, and will be given to

Table 5.2.1   Water Distribution Positions That Require an Operator's License

Job Title	Minimum License Required for Position	Time Frame to Pass Exam
Hydrant Maintenance Operator	D	For all new hires 24 months from date of hire
Maintenance Coordinator	D	For all new hires 24 months from date of hire
Distribution System Operator	D	For all new hires 24 months from date of hire
UDF Foreman	C	For all new hires 24 months from date of hire
Distribution System Technician	C	For all new hires 24 months from date of hire
Distribution System Supervisor	C	For all new hires 24 months from date of hire
Distribution System Senior Technician	B	For all new hires 24 months from date of hire
Utility Protection Coordinator	B	For all new hires 24 months from date of hire
Distribution System Engineering Manager	A	For all new hires 24 months from date of hire
Senior Distribution System Technical Manager	A	For all new hires 24 months from date of hire
Senior Distribution Supervisor	A	For all new hires 24 months from date of hire
Assistant Director of Water Distribution	A	For all new hires 24 months from date of hire

the Distribution System Support Technician to be copied and filed. The original will be returned to the Associate who will provide examination results to his/her supervisor.

Associates are encouraged to enroll in the Certification Classes taught in-house. Study guides are available from the Distribution System Support Technician. Operator Trainees are expected to use personal time to study for exams. The Company will provide on-the-job study time, as work conditions allow. On-the-job study time shall be coordinated with their supervisor. In order to continue in a position, each associate is responsible to apply for an exam, study, and pass each exam level before the specified license deadline.

The exam must be taken at an approved location. Time to travel and take the exam will be given. If after the second failed attempt taking the examination, and the deadline for examination has passed, the Associate may be terminated.

If the Operator Trainee is unsuccessful in passing the examination, then he or she must reapply to take a follow-up examination. The follow-up exam will be scheduled as soon as possible. If the examination schedule is such that a second examination would be after the specified deadline, then the Company will consider an extension to ensure that each Operator Trainee has at least two attempts to pass the examination. However, the second exam must be within the time frame allotted

---

To: *Insert Associate's Name*

From: *Insert Manager's Name, Insert Manager's Title*
           Insert Treatment Plant Name

Re: Memorandum of Understanding
           Water Distribution Licensing Program

Date: *Insert Date here*

The Water Distribution Licensing Program is designed to require all *Water Distribution Operator Trainees* to successfully complete the Certification Examination level specific to their job description and to encourage obtaining the "A" level License as well. Your specific schedule for attaining each level of required certification is as follows:

Receive passing grade for **enter level** Certification Exam before **enter date 24 months from date of hire**.

As an Operator Trainee, you are required to provide your supervisor with grade results for each Examination. In the event that only one Examination for each level is offered within the above-specified time limits, you will be allowed a time extension to enable you to take each Certification Examination level at least twice.

If you do not successfully complete the Certification Examinations with a passing grade within the specified time limits, your temporary position may end and you may be terminated from employment with the Commissioners of Public Works. Upon notification of passing each level of Certification Examination **you will be eligible for an incentive of $100.00**.

Once your Operator Trainee License is received you must have it on your person at all jobsites. You will be responsible for any and all fines if State officials come on a jobsite to inspect and you do not have it available.

I completely understand the requirements of the Operator Series as set forth by the policy and agree to those conditions.

_____   _____
Operator-in-Training Signature   Date

_____
Department Head Signature

Figure 5.2.1   Interoffice Communication Memorandum

by the LLB (Labor and Licensing Board). The Operator Trainee will bear the cost of any repeat examinations.

Licensing Incentives. Upon passing each level of examinations, the Water Distribution Operator is eligible for a $100.00 incentive (nonexempt only).

Renewals. Associates must ensure that the annual registration forms are submitted to the department administrative staff for payment. They must ensure that they have obtained the necessary continuing education units (CEUs) or contact hours to maintain their certification. Loss of certification may result in termination or transfer.

Should an associate move to another department, the associate will be responsible for renewing his or her license and paying the fee, should he or she desire to keep their Water Distribution Operator's License.

Questions to Check Progress

1. Has the utility identified proper education, training, skills test requirements, and experience as required by the local regulatory agencies for the staff conducting various actions in the distribution system for O&M?

2. Has the utility determined licensing requirements for each staff member involved in the O&M of the distribution system?

5.2.2 Competence, Awareness, and Training

The utility shall

a. Determine the necessary competence for personnel performing work affecting the distribution system.

b. Provide training or take other actions to satisfy these needs.

c. Evaluate the effectiveness of the actions taken.

d. Ensure that its personnel are aware of the relevance and importance of their activities.

e. Maintain appropriate records of education, training, skills, and experience (see 5.1.3).

Rationale

If time is taken up-front to determine what level of competence is needed by each person or position title, then the managers of a distribution system can ensure that they make the appropriate training available to their workers. In addition to just performing their job duties more effectively, when people understand what they should be doing and why their actions are important, they accept responsibility more readily, become more self-motivated, and become more committed and prideful in their work. Training records must be maintained to ensure that all workers receive the necessary levels of training determined for competency.

Example of Methods or Procedures

The following form can be used to track training topics necessary to an associate's competency.

Title: Training Matrix

Figure 5.2.2 is used to identify associates' training needs in order to provide specific training to those associates. List the title of the position on the left and mark specific training required with an "X." Gray shaded columns are reserved for specific/additional departmental training. Gray shaded column headings may be altered, and additional columns added/deleted as necessary. This form is to be completed annually and as new positions are created.

The original (record copy) shall be retained.

Associates, Positions, or Groups	Company Awareness	Company Policy	Company Procedures	Aspects/Aspects Grading Sheet	Significant Aspects	Objectives/Targets/ Improvement Plans	Legal and Other	Departmental Structure/ Responsibilities	Standard Operating Instructions Specific to Position	Departmental Emergency Plan(s)
Frequency H = At Hire, A = Annual, R = As Revised	H	H/A	H/R	H/A	H/A	H/A	H/A	H/A	H/R	H/A

Figure 5.2.2   Training Matrix

Questions to Check Progress

1. Has the utility determined the competency, awareness, and training for each position involved in the O&M of the distribution system?

2. Has the utility evaluated actual competence and awareness of the employees in a specific position for the requirements?

3. Does the utility provide vigorous training to ensure staff acquires the appropriate skills and licensing requirements?

4. Does the utility on a routine basis review the competence requirements and provide training as needed to comply with the changing requirements?

SECTION 6: REFERENCES AND RESOURCES

Allen, M.J., S.C. Edberg, and D.J. Reasoner. 2004. *Heterotrophic Plate Count Bacteria—What Is Their Significance in Drinking Water?* Awwa Research Foundation, Denver.

Ameen, J.S. 1974. *Community Water System Source Book.* Highpoint Publishing, Dripping Springs, Texas.

American Water Works Association. 1997. ANSI/AWWA Standard D102 Coating Steel Water-Storage Tanks. American Water Works Association, Denver.

———. 1998. Manual of Water Supply Practice. *M42, Steel Water-Storage Tanks.* American Water Works Association, Denver.

———. 1999. Manual of Water Supply Practice. *M6, Water Meters—Selection, Installation, Testing, and Maintenance.* 4th ed. American Water Works Association, Denver.

———. 2001. Manual of Water Supply Practice. *M2, Instrumentation & Control.* 3rd ed. American Water Works Association, Denver.

———. 2002. Manual of Water Supply Practice. *M23, PVC Pipe—Design and Installation.* 2nd ed. American Water Works Association, Denver.

———. 2003. *Principles and Practices of Water Supply Operations: Water Transmission and Distribution.* 3rd ed. American Water Works Association, Denver.

———. 2004. Manual of Water Supply Practice. *M14, Recommended Practice for Backflow Prevention and Cross-Connection Control.* 3rd ed. American Water Works Association, Denver.

———. 2004. Manual of Water Supply Practice. *M27, External Corrosion: Introduction to Chemistry and Control.* 2nd ed. American Water Works Association, Denver.

———. 2005. ANSI/AWWA Standard C651 Disinfecting Water Mains. American Water Works Association, Denver.

———. 2005. *Standard Methods for the Examination of Water and Wastewater.* 25th ed. American Water Works Association, American Public Health Association, and Water Environment Federation, Denver.

———. 2006. ANSI/AWWA Standard D102 Coating Steel Water-Storage Tanks. American Water Works Association, Denver.

———. 2006. Manual of Water Supply Practice. *M17, Installation, Field Testing, and Maintenance of Fire Hydrants.* 4th ed. American Water Works Association, Denver.

———. 2006. Manual of Water Supply Practice. *M20, Water Chlorination/Chloramination Practices and Principles*. 2nd ed. American Water Works Association, Denver.

———. 2006. Manual of Water Supply Practice. *M56, Fundamentals and Control of Nitrification in Chloraminated Drinking Water Distribution Systems*. 1st ed. American Water Works Association, Denver.

———. 2008. Manual of Water Supply Practice. *M31, Distribution System Requirements for Fire Protection*. 4th ed. American Water Works Association, Denver.

———. 2009. ANSI/AWWA Standard G200, *Distribution Systems Operation and Management*. American Water Works Association, Denver.

———. 2009. Manual of Water Supply Practice. *M36, Water Audits and Loss Control Programs*. 3rd ed. American Water Works Association, Denver.

———. 2009. *Selecting Disinfectants in a Security-Conscious Environment*. American Water Works Association, Denver.

———. 2009. Specific Taste and Odor Complaints. AWWA Professional and Technical Resources. www.awwa.org/Resources/Content.cfm?ItemNumber=585

Baribeau, Helene, et al. 2006. *Formation and Decay of Disinfection By-Products in the Distribution System*. Awwa Research Foundation, Denver.

Chadderton, R.A., et al. 1992. *Implementation and Optimization of Distribution Flushing Programs*. Awwa Research Foundation, Denver.

Kirmeyer, Gregory, et al. 2002. *Guidance Manual for Monitoring Distribution System Water Quality*. 90882. Awwa Research Foundation and American Water Works Association, Denver.

Mays, L.W. 2000. *Water Distribution Systems Handbook*. McGraw-Hill and American Water Works Association, Denver.

Montana Department of Environmental Quality Remediation Division. 2007. DEQ Technical Guidance Document #16. *Permeation of Waterlines by Petroleum Constituents*. www.deq.state.mt.us/LUST/TechGuidDocs/Techguid16.pdf

Snoeyink, Vernon L., et al. 2006. *Drinking Water Distribution Systems: Assessing and Reducing Risks*. National Academies Press, Washington, D.C.

US Environmental Protection Agency. 1991. *Lead and Copper Rule: Quick Reference Guide*. www.epa.gov/safewater/lcrmr/pdfs/qrg_lcmr_2004.pdf

———. 2003. *Cross Connection Control Manual*. www.epa.gov/OGWDW/crossconnectioncontrol/pdfs/crossconnection.pdf

———. 2008. *Ensuring a Sustainable Future: An Energy Management Guidebook for Wastewater and Water Utilities.* www.epa.gov/waterinfrastructure/pdfs/guidebook_si_energymanagement.pdf

———. 2009. Drinking Water Contaminants. www.epa.gov/safewater/contaminants/index.html

Von Huben, H., ed. 2005. *Water Distribution Operator Training Handbook.* 3rd ed. American Water Works Association, Denver.

Whelton, A.J., et al. 2004. Detecting Contaminated Drinking Water: Harnessing Consumer Complaints. *AWWA WQTC Proceedings*, Denver, Colo.

SECTION 7: AUDIT CHECKLIST

Checklist Question (AWWA Standard G200)	Remarks & Evidence	Percent Complete
4.1 Water Quality		
4.1.1 Compliance With Regulatory Requirements		
Has the utility identified and documented its legal and other requirements?		
Verify that documentation exists and is available to those who need it.		
Does the utility comply with all identified legal and other requirements?		
Verify examples of compliance, which may include reports routinely submitted to the local regulatory agency.		
If any special regulatory orders or treatment technology orders are in progress, verify compliance as per the special regulatory orders or instructions.		
4.1.2 Monitoring and Control 4.1.2.1 Sampling Plan		
Does the utility have a routine sampling plan that is representative of the entire distribution system?		
Review the plan and verify that it is representative of the system.		
Is the plan reviewed at least annually to incorporate the changes in the distribution system that may affect water quality?		
Verify procedures are in place to ensure annual sampling plan review.		

Checklist Question (AWWA Standard G200)	Remarks & Evidence	Percent Complete
4.1.2 Monitoring and Control 4.1.2.2 Sample Sites Does the utility have a minimum number of sampling sites to meet or exceed regulatory requirements? *Compare the sampling sites for the utility with applicable regulations.* Are some sites located where the longest detention time is expected, dead-end locations, areas of low circulation, water storage facilities, and where water quality problems have occurred in the past? *Interview associate(s) responsible for sample site selection and/or view procedures.*		
4.1.2 Monitoring and Control 4.1.2.3 Sample Collection Are samples collected in accordance with *Standard Methods for the Examination of Water and Wastewater*? *Review procedures for sample collection.* Does the utility utilize Chain of Custody forms and standardized labels and laboratory forms? *Review forms and interview sample collectors and/or laboratory personnel.*		
4.1.2 Monitoring and Control 4.1.2.4 Sample Taps Are sampling taps protected from outside sources of contamination? *Review the standard sample tap design.* Is the integrity of the sampling taps evaluated at least annually to correct leaks or other sources of potential contamination? *Review procedures and/or interview associate(s) responsible for sample tap inspections.*		
4.1.3 Disinfectant Residual Maintenance 4.1.3.1 Disinfectant Residual Does the utility maintain a detectable disinfection residual or a heterotrophic plate count of 500 or fewer colony-forming units per mL at all points of the distribution system at all times? *Verify that standard procedures call for a minimum disinfection residual and review documentation where sample results were recorded.*		

Checklist Question (AWWA Standard G200)	Remarks & Evidence	Percent Complete
4.1.3 Disinfectant Residual Maintenance 4.1.3.2 Nitrification Control If the utility adds ammonia or has a significant concentration in the raw water, do they monitor the free ammonia concentrations before and after the chloramination? *Verify sample records and/or review procedures.* Are disinfection goals based on historical data to avoid nitrification? *Interview key personnel responsible for setting disinfection goals.* Are the key indicators for nitrification parameters monitored, i.e., nitrite, nitrate, free ammonia, etc.? *View monitoring records and/or reports.*		
4.1.3 Disinfectant Residual Maintenance 4.1.3.3 Booster Disinfection 4.1.3.3.1 Residual Goals Documentation 4.1.3.3.2 Operating Procedures 4.1.3.3.3 Written Response Plan Verify: If the utility uses a residual disinfectant and booster disinfection, have residual goals been established and documented, as well as a means to monitor compliance with those goals? *View goals and monitoring records.* Are there written operating procedures for each booster station that take into account seasonal variations, water quality, flow, and system operation variations? *View operating procedures.* Are there written plans to respond when measured results do not meet operational goals? *View procedures/plans.*		
4.1.3 Disinfectant Residual Maintenance 4.1.3.4 Disinfection By-product Monitoring & Control 4.1.3.4.1 Program to Control & Monitor By-Products 4.1.3.4.2 Action Plan to Respond Does the utility have a monitoring program for the control of disinfection by-products, including specified goals and actions necessary to respond to DBP problems in the distribution system? *Review the procedures that make up the DBP monitoring and control program for the above criteria.*		

Checklist Question (AWWA Standard G200)	Remarks & Evidence	Percent Complete
4.1.4 Additional Requirements for Utilities Not Utilizing a Disinfectant Residual 4.1.4.1 Response Program 1. Goals for HPC 2. Criteria for Initiation of Defined Actions 3. Criteria for Initiation of Corrective Actions 4. Description of Responsibilities Does the utility monitor and record HPCs? *View records and/or reports.* Has the utility established an action plan that includes goals for HPCs at the critical points in the distribution system and criteria for initiation of response actions to correct problems before they become a health concern? *Review procedures that make up the HPC action plan for the above criteria.*		
4.1.5 Internal Corrosion Monitoring and Control 4.1.5.1 Prevention and Response Program 1. Monitoring and Sampling Plan 2. Inspection of Pipe Conditions 3. Procedures for Controlling Lead and Copper 4. Guidelines for Controlling of Corrosion Related By-Products Has the utility established a monitoring and control plan to respond to internal corrosion and disposition problems in the distribution system? Does this plan call for the monitoring and sampling of pH, alkalinity, conductivity, phosphates, silicates, calcium, metals, asbestos, etc.; inspection of exposed pipelines for condition, structural integrity, and hydraulic capacity; procedures to control lead and copper; and guidelines to control other parameters, such as iron, zinc, color, and taste and odor? *Review procedures related to the corrosion control plan.*		

Checklist Question (AWWA Standard G200)	Remarks & Evidence	Percent Complete
4.1.6 Aesthetic Water Quality Parameters 4.1.6.1 Color and Staining 1. Inquiry Call System 2. Trained Response Personnel 3. Communication of Inquiry Is an inquiry call system that identifies and tracks color and staining problems in place? *View call log/system for verification.* Are personnel trained to handle these types of inquiries? *View training records, procedures, and/or interview personnel.* Are these inquiries communicated effectively and are records reviewed periodically to identify problem areas of the system? *View records and/or interview personnel.*		
4.1.6 Aesthetic Water Quality Parameters 4.1.6.2 Taste and Odor 1. Inquiry Call System 2. Trained Response Personnel 3. Communication of Inquiry Is an inquiry call system that identifies and tracks taste and odor concerns in place? *View call logs/system for verification.* Are personnel trained to handle these types of inquiries? *View training records, procedures, and/or interview personnel.* Are these inquiries communicated effectively for follow-up and are records reviewed periodically to identify problem areas of the system? *View records and/or interview personnel.*		

Checklist Question (AWWA Standard G200)	Remarks & Evidence	Percent Complete
4.1.7 Customer Relations 4.1.7.1 Customer Inquiries 1. Recorded Customer Identification 2. Documented Number of Calls 3. Reduction of Calls Goal 4. Documented Initial Contact 5. Inquiry Evaluation 6. Trained Interview Personnel 7. Timely Response 8. Problem Area Identified Is a system in place to document all customer inquiries? Does the system record the customer's identification, specific inquiry time, result of investigation, and resolution of the inquiry; include the number of inquiries and annual goal to reduce number of water quality complaints; and track trends? *Review customer inquiry system for the above criteria.* Are trends reviewed to identify problem areas in the distribution system? *View documentation and/or interview personnel.*		
4.1.7 Customer Relations 4.1.7.2 Service Interruptions Does the utility have a system for documenting all planned and unplanned service interruptions? *View the system.* Does the utility establish an annual goal to reduce these interruptions? *View documentation, reports, and/or records.*		
4.1.8 System Flushing 1. Preventive Program 2. Flushing at Appropriate Velocity 3. Written Procedures for all Flushing Activities Does the utility use a systemwide integrated flushing program that takes a preventive approach to flushing, including spot flushing for isolated water quality concerns and routine flushing to avoid problems? Are all the procedures associated with the flushing program documented? *View procedures for the flushing program.*		

Checklist Question (AWWA Standard G200)	Remarks & Evidence	Percent Complete
4.2 Distribution System Management Programs		
4.2.1 System Pressure 4.2.1.1 Minimum Residual Pressure 4.2.1.2 Pressure Monitoring Are procedures in place to ensure that residual pressure is maintained above 20 psi? Is the pressure monitored with alarms in place to alert operators of any pressure conditions outside the utility's requirements? *Review procedures referencing a minimum residual pressure, any monitoring reports, and/or SCADA system set points.*		
4.2.2 Backflow Prevention Does the utility have a comprehensive cross-connection and backflow prevention program modeled after AWWA Manual M14? *Review procedures that make up the backflow prevention program.*		
4.2.3 Permeation Prevention Are there procedures in place to help prevent permeation of organic solvents into the system through PVC piping, seals, etc.? *Review any procedures or standards in place for permeation prevention.*		
4.2.4 Water Losses 4.2.4.1 Water Loss Does the utility have an annual goal for unaccounted-for water loss and a standard formulation for calculation? *View any procedures, forms, and/or records related to the determination of unaccounted-for water.*		
4.2.4 Water Losses 4.2.4.2 Response Program Is there a formal action plan in place for responding to a situation where unaccounted-for water exceeds the annual goal set by the utility? *Review any documentation related to a response program for excessive unaccounted-for water loss.*		
4.2.4 Water Losses 4.2.4.3 Leakage Does the utility have a method for estimating leakage from the distribution system? Is it expressed in gal/day/mi of piping? *Review reports and/or procedures associated with distribution system leakage.*		

Checklist Question (AWWA Standard G200)	Remarks & Evidence	Percent Complete
4.2.5 Valve Exercising and Replacement 4.2.5.1 Valve Exercising Program Does the utility have a valve exercising program in place that covers the following elements: • A goal for the number of transmission valves to be exercised annually? • A goal for the number of distribution valves to be exercised annually? • Means to verify goals are being met? • An action plan in the event that goals are not met? • Identification of critical valves? • Goals to reduce the percentage of inoperable valves in the system? *Review the procedures that make up the valve exercising program for the above elements.*		
4.2.6 Fire Hydrant Maintenance and Testing 4.2.6.1 Maintenance and Testing Does the utility have a hydrant testing and maintenance program that covers the following elements: 1. A goal for the number of hydrants to be inspected and tested annually? 2. Procedures for properly opening and closing hydrants? 3. Fire flow test requirements? *Review the procedures that make up the fire hydrant maintenance and testing program for the above elements.*		
4.2.7 Material in Contact With Potable Water 4.2.7.1 Approved Coating or Linings Does the utility ensure that only NSF/ANSI Standard 61 approved coatings and lining are used throughout the distribution system? *Review procedures and/or specification that require only NSF/ANSI Standard 61 coatings and linings.*		
4.2.8 Metering 4.2.8.1 Metering Requirements Does the utility monitor the treated water entering the distribution system to determine daily peak flows and maximum-day peak flows? *Review records and/or reports showing the metered influent to the distribution system.*		

Checklist Question (AWWA Standard G200)	Remarks & Evidence	Percent Complete
4.2.8 Metering 4.2.8.2 Metering Devices Do all metering devices meet AWWA requirements? *Review specifications and/or interview key personnel.*		
4.2.8 Metering 4.2.8.3 Testing Does the utility test and replace meters per AWWA Manual M6 standards? *Review program procedures and goals and/or interview key personnel.*		
4.2.8 Metering 4.2.8.4 Repair and Replacement Programs Does the utility have a program to repair and replace defective meters? Does the program conform to the guidelines set forth in AWWA Manual M6? *Review records and procedures associated with meter repair and replacement.*		
4.2.9 Flow 4.2.9.1 Flow Requirements Is the system capable of delivering maximum-day demand and fire flow requirements? *View hydraulic model results, reports, and/or interview key personnel.*		
4.2.10 External Corrosion 4.2.10.1 Leaks/Breaks Does the utility record information related to main breaks including, at a minimum, location, pipe material, size, type of break, soil type, pipe depth, and soil saturation conditions prior to break? *View records and/or policy/procedures for obtaining and recording main break information.*		
4.2.10 External Corrosion 4.2.10.2 Monitoring Program Does the utility have an external corrosion monitoring plan, such as potential measurements, line current measurements, soil resistivity, and soil chemical analysis? *Review any procedures that make up the external corrosion monitoring program.*		

Checklist Question (AWWA Standard G200)	Remarks & Evidence	Percent Complete
4.2.11 Design Review for Water Quality 4.2.11.1 Policies and Procedures Does the utility comprehensively review all construction projects for potential water quality degradation before and after installation? *Review policies/procedures for the review of construction projects.*		
4.2.11 Design Review for Water Quality 4.2.11.2 Records Does the utility maintain as-built records of all newly installed and retrofitted facilities? *View filing system for as-builds and/or interview key personnel.*		
4.2.12 Energy Management 4.2.12.1 Energy Management Program Does the utility have an energy management program in place that identifies energy usage and trends and takes into account energy costs when planning new distribution facilities? *View energy inventory/audit records, procedures that are part of an energy management program, and/or interview key personnel.*		
4.3 Facility Operations and Maintenance		
4.3.1 Treated Water Storage Facilities 4.3.1.1 Storage Capacity Are minimum operating levels in the storage facilities based on pressures, firm flow requirements, emergency storage, and other site-specific requirements? *Review any reports, calculations, hydraulic models, and/or interview key personnel.*		
4.3.1 Treated Water Storage Facilities 4.3.1.2 Operating Procedures Are there written operating procedures that address water level fluctuations and turnover rates, with a target turnover rate? *View the procedures for water storage operation.*		
4.3.1 Treated Water Storage Facilities 4.3.1.3 Inspection Is there a written inspection program for water storage facilities? Does the program outline frequency (routine, periodic, and comprehensive), procedures, and record keeping? *Review records and procedures for treated water storage inspections.*		

Checklist Question (AWWA Standard G200)	Remarks & Evidence	Percent Complete
4.3.1 Treated Water Storage Facilities 4.3.1.4 Maintenance Does the utility have a formal maintenance program outlined for treated water storage facilities? Does the program include periodic cleaning and refurbishing as needed? *Review the procedures that are part of the treated water storage maintenance program.*		
4.3.1 Treated Water Storage Facilities 4.3.1.5 Disinfection Does the utility have a formal policy for the disinfection of treated water storage facilities prior to placing them on-line? Does the policy follow ANSI/AWWA Standard C652, and address the disposal of heavily chlorinated water into the environment? *Review the procedures for treated water storage facilities disinfection.*		
4.3.1 Treated Water Storage Facilities 4.3.1.6 Additional Requirements Are all treated water storage facilities covered and protected from contamination, or provided with additional treatment? *Tour facilities, review site plans/drawings, and/or interview key personnel.*		
4.3.2 Pump Station Operations and Maintenance 4.3.2.1 Operating Procedures Does the utility have written operating procedures for each pump station and record operational conditions, such as inlet pressures, flow rates, etc.? *View operating procedures and operational logs.*		
4.3.2 Pump Station Operations and Maintenance 4.3.2.2 Maintenance Program Does the utility have written maintenance procedures for each pump station and maintain records to document service performed? *View maintenance procedures and records.*		

Checklist Question (AWWA Standard G200)	Remarks & Evidence	Percent Complete
4.3.3 Pipeline Rehabilitation and Replacement 4.3.3.1 Rehabilitation and Replacement Program Does the utility keep the following records on installed pipelines to help with condition assessment? 1. Current Maps 2. Maintenance Records 3. System Data 4. Environmental Information 5. Proper Separation 6. Annual Goal for Breaks *View records and/or database information.*		
4.3.4 Disinfection of New or Repaired Pipes 4.3.4.1 Disinfection of New or Repaired Pipes Are all new or repaired pipes disinfected in accordance with ANSI/AWWA Standard C651 prior to being placed in service? *View procedures and/or interview key personnel.*		
4.3.4 Disinfection of New or Repaired Pipes 4.3.4.2 Bacteriological Testing Is bacteriological testing completed in accordance with ANSI/AWWA Standard C651? *View procedure for Bac-T testing.*		
4.3.4 Disinfection of New or Repaired Pipes 4.3.4.3 Disposal of Chlorinated Water Is the disposal of heavily chlorinated water done in accordance with all applicable regulations? *View procedures and interview key personnel.*		
5.1 Documentation Required **5.1.1 General** Does the documentation for the elements of Sec. 4 consist of the following: • Quality Policy and Objectives • Standard Operating Procedures • Documented Procedures • Planning Documents • Required Records *Verify documentation throughout the entire audit process.*		

Checklist Question (AWWA Standard G200)	Remarks & Evidence	Percent Complete
5.1.2 Examples of Documentation (N/A)		
5.1.3 Control of Documents		
Does the utility have a standard procedure for the approval of documents prior to use?		
Is there a procedure in place to approve changes to documents and ensure that the most up-to-date version is available?		
Are documents of external origin identified and controlled?		
Does the utility destroy obsolete documents or clearly identify them to avoid unintended use?		
View document control procedures and/or interview key personnel.		
5.1.4 Control of Records		
Has the utility established procedures for the identification, storage, protection, retrieval, retention time, and disposition of records?		
Review record retention schedule and/or record control procedures.		
5.2 Human Resources		
5.2.1 General		
5.2.2 Competence, Awareness, and Training		
Has the utility determined the level of competency (training requirements, licenses, etc.) required for personnel working on the distribution system?		
Is such training provided and records maintained?		
Review training matrixes, schedules, and/or plans. Compare training matrixes to records.